PMP® Exam

MINDSET

*The servant-leader success mindset for the PMP® Exam
Based on the PMP® Exam Content Outline, PMBOK® Guide
Sixth Edition, Seventh Edition, PMP® Exam Content Outline,
Agile Practice Guide and the Agile Manifesto*

PHILL AKINWALE, OPM3, PMP, PMI-ACP

PraizionMedia
Real World Project Management Training Solutions

PMP® Exam Mindset

Published by Praizion Media

P.O Box 22241, Mesa, AZ 85277

E-mail: info@praizion.com

www.praizion.com

Author

Phillip Akinwale, MSc, OPM3, PMP, PMI-ACP, CAPM, PSM, CSM

ISBN 978-1-934579-92-3

ISBN 9781934579923

9 781934 579923

CONTENTS

INTRODUCTION

I have students who are taking the PMP® exam, and the question that I often get is "What do I need to do when it is one, two or three days to my PMP® exam?"

The answer is: solidify the PMP® Exam MINDSET!

In this book we're going to focus on the right mindset and agile thinking for your PMP® exam in people, process, and business. I will also give you some "one day to the PMP® exam" advice, all rolled up in one.

This is a very condensed book where I will squeeze hours of learning into minutes.

PART 1: The PMP® Exam Success MINDSET Mantras

CHAPTER ONE: PEOPLE PERFORMANCE DOMAIN

I have broken this down into People, Process and Business. Let's talk about the people mindset. For the exam, you've got to be familiar and ready to see words such as team, customers, stakeholders, sponsor, product owner, program manager, and so on.

Imagine the greatest orchestras in the world. They have the finest instruments and the grandest concert halls, but without the harmony and cohesion between the musicians, they could never deliver beautiful symphonies. The same principle applies to business. Tools, techniques, and strategies are important, but it's the people who bring projects to life, who drive innovation, and who ultimately lead an organization to success. This is where the People Mindset comes into play.

The People Mindset is all about putting people at the heart of everything we do. It's about fostering a culture where every individual feels valued and heard, where their ideas matter, and where they are empowered to grow and contribute to their fullest potential.

As we explore this section, we will study key facets of the People Mindset, from customer obsession and team focus to integrity, fairness, trust, and beyond. Through real-life anecdotes, inspirational quotes, and practical examples, we aim to not just impart knowledge but also inspire a shift in perspective that will empower you to bring out the best in your team and stakeholders.

As Richard Branson, founder of the Virgin Group, once said, "A company is people... employees want to know... am I being listened to or am I a cog in the wheel? People really need to feel wanted." This encapsulates the essence of a People Mindset - understanding that the most valuable asset of any organization is its people. So let's dive in and explore how we can truly put people first.

1.1 Customer Obsession: Make your customer's success a primary goal

In any project, the customer should be the focal point of all decision-making processes. This involves understanding the customer's needs and expectations and prioritizing them in the project's strategy and execution.

Have you ever heard the phrase, "Customer is king"? Well, it's not just a catchy slogan; it should be our mantra. Let's put the customer at the heart of every decision we make, every strategy we devise. We need to dig deep, understand what our customers want, what they need, and then deliver it to them in the best possible way. As Jeff Bezos once said, "We see our customers as invited guests to a party, and we are the hosts. It's our job every day to make every important aspect of the customer experience a little bit better."

The first thing you want to be thinking about is the customer. Be obsessed with the customer, make your customers' success a primary goal. Always look out for the customer.

Customer focus involves identifying all stakeholders, understanding their interests and expectations, and managing their engagement throughout the project. Stakeholder care includes regular communication, ensuring their needs are met, and incorporating their feedback into the project.

Our stakeholders are partners in our journey. We need to identify all of them, understand their needs and expectations, and ensure they feel involved and valued throughout the project. It's about regular communication, addressing their concerns, and considering their input. As Peter Drucker said, "The aim of marketing is to know and understand the customer so well the product or service fits him and sells itself."

1.2 Advancement and Project Progress: Work with the customer to move the project forward

Maintaining a focus on continuous improvement and advancement of the project. This involves regular tracking of project progress, identifying areas of improvement, and implementing changes that enhance efficiency and productivity.

Let's be like sharks: if we stop moving, we stop progressing. We need to continuously track our project's progress, identify areas for improvement, and implement changes that enhance our productivity and efficiency. As Mark Zuckerberg said, "The biggest risk is not taking any risk... In a world that's changing really quickly, the only strategy that is guaranteed to fail is not taking risks."

Always move the project forward, I cannot overstate this. This is by far one of the most important points for your exam. Always choose the option that moves the project forward.

Let me give you an example, you get a question that states "You are a project manager with a stakeholder who has refused to sign documentation, what should you do? And you are presented with the following options:

a)   Report that stakeholder to their boss.

•   What good does that do in the grand scheme?

b)   b) discuss with the stakeholder and understand their point of view.

•   That is the better option.

So any option that does not move the problem or the issue forward or advance the project, do not choose! Always choose the best option to advance the project forward.

1.3 Fairness and Empathy: Treat others fairly, have empathy, be mindful and aware of diversity and inclusion.

A people-centric mindset involves treating all individuals fairly and with empathy. This includes understanding and respecting diversity, fostering an inclusive environment, and being empathetic to the needs and challenges of team members and stakeholders.

Every team member is unique and adds value. We need to respect their individuality, understand their challenges, and treat everyone fairly. As Atticus Finch said in 'To Kill a Mockingbird', "You never really understand a person until you consider things from his point of view."

Treat others fairly, have empathy, be mindful and aware of diversity and inclusion. Assume you're a project manager on a project and there's a team member that has been ostracized in some way by the other team members, what should you do? Should you just allow that team member to be all on their own? You find out that they're not including this person in discussions, in debate, you're realizing that the person is always isolated and they're not including the person in group activities. What should you do as a scrum master, as a servant leader, as a project manager?

Remember, you are a coach, a mentor, so you need to mentor and coach these individuals into inclusion and diversity. That is what you as a leader needs to be thinking about. Treat others fairly, have empathy, be mindful and aware of diversity, and also that everyone is included.

1.4 Stewardship of Resources: Protect resources entrusted to you & treat with care (fiduciary)

Be a good steward! Protect resources and trust entrusted to you and treat them with care. Remember you are a fiduciary of these resources, so treat team members with care, treat physical resources with care, ensure that they are well-taken care of and allocated as they are supposed to be. Do not abuse the resources that are in your care.

Responsible management and utilization of all project resources. This includes careful budgeting, efficient use of materials, and mindful allocation of human resources to maximize project outcomes.

We are the caretakers of our project resources. This means we need to budget carefully, use materials wisely, and deploy our human resources strategically to maximize the project outcomes. As Antoine de Saint-Exupery wrote, "A designer knows he has achieved perfection not when there is nothing left to add, but when there is nothing left to take away."

1.5 Focus on the Team: Focus on stakeholder & team health, well-being and synergy

The team is the backbone of any project. A people-centric mindset involves focusing on building a cohesive, skilled, and motivated team. This includes facilitating effective communication, resolving conflicts, and providing the necessary resources for the team to execute their tasks successfully.

Remember, a project is only as good as the team behind it. Our team is our strength, our superpower. Building a strong, cohesive, and motivated team is the key to our success. We should foster open communication, resolve conflicts swiftly and fairly, and provide the resources our team needs to excel. As Phil Jackson wisely put it, "The strength of the team is each individual member. The strength of each member is the team."

Focus on the team and their well-being and their health and their synergy.

1.6 Trust: Trust the team and their judgment. Allow them choose their way of working

Trust the team and their judgment. The questions will test you on understanding that the team should be self-organizing, autonomous, and you should believe the best in the team. Allow the team to choose their own way of working.

Trust and honesty form the foundation of effective team dynamics. This includes being honest in all communications, meeting commitments, and building an environment where team members feel safe to express their ideas and concerns.

Trust and honesty are like glue holding our team together. We need to foster an environment of openness and reliability where everyone feels secure to share their ideas and concerns. As Stephen Covey advised, "Trust is the glue of life. It's the most essential ingredient in effective communication."

1.7 Servant Leadership: Defend the team, be a diversion shield and facilitate conflict resolution

The summary of this mantra is: defend the team, be a diversion shield and facilitate conflict resolution. What we're talking about here is to protect the team to act in a capacity of diverting anything coming at the team from a distracting point of view, people wanted to throw more work on the team that has nothing to do with the ongoing project or ongoing endeavor.

You should act as a diversion shield. You should also act as a shield to remove impediments, obstacles, blockers, you should use your network as a project manager or as a servant leader to find ways of removing any blockers around. Key feedback we've heard from a lot of students who take the exam is that, the word servant leader may not be used as much but you can tell it is a servant leader being discussed. You can also tell that the word scrub master may not show up as much, but you know this is talking about a servant leader figure so get comfortable with the idea that they could still come at you from the angle of project manager, but you need to be thinking servant leader.

Adopting a servant leadership approach where the leader prioritizes the needs of the team and helps them perform at their best. This includes providing guidance, removing obstacles, and empowering team members to take ownership of their tasks.

As leaders, let's serve first and lead second. It's about providing guidance, removing obstacles, and empowering team members to take ownership of their tasks. Robert Greenleaf, the father of servant leadership, wrote, "The servant-leader is servant first... It begins with the natural feeling that one wants to serve, to serve first."

1.8 Mentorship and Coaching: Mentor, coach, serve and guide the team instead of using punishment and coercion

Mentor coach, serve, and guide the team instead of using punishment and coercion. Your go-to response needs to be one of mentoring and coaching in order to get the team on the same page to introduce the behaviors you desire and things like that.

Providing guidance, mentorship, and coaching to team members to enhance their skills and performance. This involves sharing knowledge, providing constructive feedback, and supporting personal and professional growth.

We're not just team leaders; we're also mentors and coaches. Our role is to help our team members develop, both professionally and personally. It's about sharing knowledge, providing constructive feedback, and supporting their growth. As John C. Maxwell said, "One of the greatest values of mentors is the ability to see ahead what others cannot see and to help them navigate a course to their destination."

1.9 Integrity and Fairness: Do not abuse your position or title or be partial in your actions

Maintaining integrity and fairness in all project activities is paramount. This involves upholding ethical standards, treating all team members and stakeholders with respect, and ensuring transparency in decision-making processes.

Think of integrity and fairness as our compass guiding us through our project journey. This involves sticking to our moral and ethical principles, treating everyone with respect, and ensuring our decision-making processes are transparent. As Warren Buffett advised, "It takes 20 years to build a reputation and five minutes to ruin it. If you think about that, you'll do things differently."

As a leader, do not abuse your position or title or be partial in your actions. Make sure you choose those options where you are always taking the higher ground where you are above board and everything you are doing is on point.

1.10 Honesty: Be honest and truthful in all your dealings even IF it may offend others

Honesty forms the foundation of effective team dynamics. This includes being honest in all communications, meeting commitments, and building an environment where team members feel safe to express their ideas and concerns.

Trust and honesty are like glue holding our team together. We need to foster an environment of openness and reliability where everyone feels secure to share their ideas and concerns. As Stephen Covey advised, "Trust is the glue of life. It's the most essential ingredient in effective communication."

Honesty, the cornerstone of integrity, encompasses a profound commitment to truthfulness and transparency in every interaction and circumstance. It demands unwavering courage, even in the face of potential offense or discomfort.

In the realm of project management, honesty plays a pivotal role in establishing trust and fostering open communication. As a project manager, your ethical compass should guide you to prioritize truth over appeasement, even if it means broaching sensitive topics or delivering unwelcome news.

While this may initially cause discomfort or resistance, the long-term benefits far outweigh any temporary unease.

By upholding honesty as a core value, you create an environment of authenticity and credibility. Stakeholders, team members, and colleagues rely on your integrity to make informed decisions and understand the reality of the project's status. Your commitment to truthfulness instills confidence in your leadership, as others recognize your genuine dedication to the project's success.

It is essential to remember that honesty does not equate to being tactless or insensitive. Honesty can be delivered with empathy and sensitivity, considering the context and the individuals involved. By choosing your words carefully, demonstrating empathy, and actively listening, you can navigate difficult conversations while still upholding the principles of honesty and respect.

Furthermore, honesty extends beyond your interactions with others; it also encompasses self-honesty. Reflecting on your own strengths and weaknesses, acknowledging mistakes, and taking responsibility for them is vital for personal and professional growth. Embracing self-honesty allows you to learn from your experiences, adapt your approach, and continuously improve as a project manager.

However, it is crucial to recognize that honesty must always be balanced with ethical considerations and professional judgment. While honesty is essential, it should not be used as an excuse for unnecessary bluntness or tactlessness. It is essential to exercise discernment and tactfully navigate situations where delivering the truth may require finesse or a thoughtful approach.

Finally, embodying honesty as a project manager empowers you to establish trust, foster open communication, and uphold your integrity. It requires the courage to speak the truth, even when it may cause discomfort or offense. By embodying honesty with empathy and self-reflection, you cultivate an environment of authenticity, credibility, and growth, ultimately contributing to the success of your projects and the trust placed in your leadership.

1.11 Leadership: Be courageous to lead. make tough decisions have tough conversations make trade-offs

This involves leading with courage and confidence, especially in the face of challenges or uncertainty. Courageous leadership includes making tough decisions, taking calculated risks, and standing up for the team when necessary.

Leading with courage and confidence, especially when faced with uncertainty, is essential. Courageous leadership means making tough decisions, taking calculated risks, and standing up for our team when necessary. As Winston Churchill said, "Courage is what it takes to stand up and speak; courage is also what it takes to sit down and listen."

Be courageous to lead, make tough decisions, have tough conversations, make trade-offs. It will amaze you that some of the questions may push you to the limit where the best option may be to let someone go from a project. You got to be bold to make those decisions.

Other instances it would not be fair to let that person go because that person is a capable individual, would be a good one to have on the team but you need to know when to use mentoring and coaching and training but they're going to be some decisions where you need to make the tough decisions. It will push you to that limit, that place you don't want to go, that place you don't want to address, it will be in your face. Trust me, it's happened to me on the exam. I have seen questions come at me right there in my face and you need to decide, make the tough decisions.

1.12 Agility: Be agile and adapt to be resilient

Embracing an agile mindset that is flexible and responsive to change. This includes being open to new ideas, adjusting plans based on feedback or changing circumstances, and fostering a culture of adaptability within the team.

Change is inevitable, and our ability to adapt is vital. We need to be flexible, open to new ideas, and ready to adjust our plans based on feedback or changing circumstances. We should foster a culture of adaptability within the team. As Leon C. Megginson wrote, "It is not the strongest of the species that survives, nor the most intelligent; it is the one most adaptable to change."

Be agile and adapt to be resilient. That is the final one in the people domain. When we talk about Agile, what are we saying? Adapt, be resilient. Even if you are in a Hybrid situation, your thinking should be Agile not Predictive, right? Because Agile means you could be more iterative or incremental or less so agility is always the best way, being Predictive is just a one-way street, you got to remember that. The agile person could always change. The predictive person will be sticking to their guns. You got to remember that.

CHAPTER TWO: PROCESS PERFORMANCE DOMAIN

et's go into the next mantra area. The second area, is a set of **Process Mantras**. When you're in the exam, you're going to face knowledge areas, process groups, processes, formulas, methods, models, artifacts, theory, frameworks, and practices and it's for this reason that I tell people to a fair amount of theory and predictive processes.

Now for those of you who have your exam coming up in the next week, let me give you some solid advice, please do this. Open up your PMBOK® Guide sixth edition. Don't go to the main body because it's already too late at this point. What I want you to do is, first of all, look for all of my videos on

YouTube where I have the one-minute summaries, so I have a one-minute summary for all of the knowledge areas and I have a summary of about five minutes or so for chapters one, two, and three. That's all you need at this point because forget it, if your exam is next week, spend some time doing other things not reading this whole book, okay? If you're good at speed reading, I want to give you a cadence here. I want you to spend on every chapter not more than 10 minutes to go through it.

I'm not talking about skimming through but I'm talking about speed power reading through. And if you want to know a little bit more about how to do this, there's some courses online where you can quickly learn tips and tricks to help you speed read. One of such individuals is Jimmy Kwik. Check out his information. He specializes in a lot of brain optimization stuff. Check out Jimmy Kwik. You'll find him on YouTube and he talks about how you can speed read, and just listen to what he has to say and apply it. But I would say, the cadence is roughly 10 minutes per PMBOK® Guide Chapter.

Don't spend any more than that and you'll be amazed after listening to Jimmy and applying that stuff to the book what you're going to gain, all right?

Now, when you have done that, if you can even do that, it will take you about two hours, so if you spend 10 minutes on each one 10 times for 13 chapters, that's 120, that's roughly two hours plus. Don't spend more than that because it's just going to be the law of diminishing returns. You're going to get all bothered about ITTOs that you're not able to memorize, don't memorize them, understand. As you're going through, just say do I understand it?

Let me give you an example, quick example. If I randomly open the PMBOK® Guide Sixth Edition to page 434 and I see a diamond then it says sensitivity analysis. I need to know what this is at a high level, that's enough, okay? For most of these, I want to say about 95% of the stuff here, if you understand it at a decent level, high level, that's enough. You don't need to go cramming every single ITTO where it comes from, no.

You don't have to. Let me tell you why. I do a lot of listening to students who take the exam. My students, other students and you know what I've heard since 2021? Feedback from students that was a broken record! I kept hearing the PMBOK® Guide and all the processes, process groups, knowledge areas, they were not verbatim on the exam, so cramming is not going to help you!

A lot of people say not even a single ITTO spelt as it is in the guide appeared on the exam. You understand what I'm saying? So be smart and go through the content.

So back to Page 434 in the PMBOK® Guide Sixth Edition, as I'm looking at sensitivity analysis here, I need to get the overall gist! Sensitivity analysis helps to determine which individual project risks or other sources of uncertainty have the most potential impact on project outcomes. boom! It's not rocket science. I don't need to go off on a rabbit trail, fretting about the image under it. The image under it is the Tornado diagram by the way.

Here's my logic, before you know, the PMBOK sixth "puritans" come at me, no one is more passionate about the sixth than I am, but something else I'm passionate about is use of time as a resource! GOOD use of time, and being lean and mean with your time. Don't waste time because no one's got time to waste. I don't. I hate people wasting my time and I want to spend yours wisely with you. So, spend two hours if you've got it on this book. I would say spend one-hour speed reading again through it.

Now when it comes to the PMBOK® Guide Seventh Edition and the Sixth Edition, one more thing I want you to do within 30 minutes is to go through the glossary items, looking for terms that you don't know. If your exam is in a week, spend the day doing what I've told you to do here and that's enough, okay? The process piece, the language maybe more technical or I should say more process-oriented, more towards the spectrum of frameworks, and that's okay. If you do what I've asked you to do with those books, you're going to be in great shape for your exam!

All right, so the Process Mantra, like I said knowledge areas, process groups, cost, scope, schedule, risk, procurement, stakeholders, all that stuff you're going to find language like this. In this section, we examine the critical mindset and approach required to effectively navigate the intricacies of project processes. As a project manager, having a strong process mindset is essential for ensuring project success from initiation to closure.

This section focuses on developing the necessary mindset to understand, implement, and optimize project processes. It equips you with the knowledge and tools to approach project management from a structured and systematic standpoint. By embracing the process mindset, you'll be able to streamline workflows, manage resources efficiently, and achieve project objectives with confidence.

Throughout this section, we'll explore key topics associated with project initiation, planning, execution, monitoring and control, and project closure. Each topic will provide valuable insights, techniques, and best practices to enhance your process-oriented thinking and decision-making.

Get ready to expand your understanding of project processes, learn effective strategies for managing project tasks, and develop the ability to optimize project outcomes through systematic approaches. You'll gain the skills to identify and mitigate risks, manage project communications, lead teams, and adapt to changing project dynamics.

Remember, a solid process mindset is the foundation of successful project management. It empowers you to navigate complex projects with clarity, agility, and efficiency.

So let us begin in earnest, the Process Mindset and embark on a journey of mastering the mindset and practices that drive project success through effective process execution.

Get ready to embrace the power of processes and elevate your project management prowess!

2.1 Life Cycle: Select & tailor the appropriate project life-cycle & development approach

The first manta in the Process section is, Life Cycle. Select and tailor the life cycle and the development approach. Understand the project life cycle is a collection of phases and those phases are peculiar to the technical work being done but most importantly the development approach must also be appropriate. Will it be iterative, incremental, predictive, or agile? So choose the right development approach!

What if I have a simple project? What if I have chaos? What if I have anarchy? Which one do I choose? If I have anarchy, I'm going to be on the side of Agile. If I've got very Predictive, simple, I'm going to be in Predictive space.

If I've got a high degree of change and a high degree of delivery, delivering frequently and rapidly, I'm going to be Agile. But if I've got a high frequency of delivery and a low degree of change, I'm going to be incremental. If I've got a low degree of change and a low degree of delivery, I'm going to be Predictive. If I've got a low degree of delivery, a low frequency of delivery and a high degree of change, I'm going to be iterative. So, you got to know pages 18 and 19 in the

Agile Practice Guide.

Understanding the nature of the project and selecting the right project life cycle is paramount. This process involves considering the technical work being done and choosing the phases that align with the project's requirements. The development approach is tailored to fit these requirements, with considerations given to iterative, incremental, predictive, or agile approaches based on the project's complexity and needs.

Remember when we've talked about the Agile principle of "Simplicity--the art of maximizing the amount of work not done--is essential"? This is crucial here. Choose the life cycle that suits your project like a tailored suit, fits perfectly without unnecessary frills. It's not a one-size-fits-all scenario, and we adapt our approach to the specific needs and complexities of each project.

2.2 Hybridize: Hybridize where necessary to maximize value and option

Hybridization in project management refers to the practice of combining elements of both Agile and predictive (traditional) project management practices. It involves tailoring the project management approach to fit the specific needs and characteristics of the project.

Hybridize where necessary to maximize value and options. Why would you hybridize? Because there's always an opportunity to be iterative in work cycles or incremental in delivery. Be open to combine the right measure of iterative and incremental practices, So hybridize when necessary. If there's an opportunity to hybridize instead of being just Predictive then hybridize!

Inspect and adapt. It's one of the tenets of Agile. We continuously inspect, adapt, and integrate on all levels. Scrum pillars, transparency, inspection, adaptation is very important. In our daily scrum, we are actually inspecting the work, we are actually adapting as well.

Be aware that practices such as a retrospective may not be

explicitly called that, but the idea of what you're doing in the retrospective is important.

In hybrid approaches, certain aspects of the project may be managed using traditional predictive techniques, while others are handled using Agile approaches. This blending of practices allows teams to leverage the strengths of both approaches and adapt to the unique requirements of the project.

The decision to hybridize is driven by the recognition that not all projects or organizations can fully adopt Agile or predictive approaches alone. By hybridizing, project teams can take advantage of the structure and planning inherent in predictive practices, while also incorporating Agile principles such as flexibility, iterative development, and customer collaboration.

Hybridization recognizes that different projects may require different levels of predictability, control, and adaptability. It allows organizations to find the right balance between structure and agility, based on factors such as project complexity, stakeholder expectations, and industry regulations.

Overall, hybridization is a strategic approach that aims to maximize value and optimize project outcomes by selectively integrating Agile and predictive practices.

2.3 Agile Mindset: Seek to deliver incrementally, plan iteratively where possible

Apply an Agile mindset! Seek to deliver incrementally, plan iteratively where possible, break the work into chunks, break delivery into chunks, break planning into chunks. What does it do? It's a risk coping mechanism. When you deliver in increments, you are narrowing down any issues that could potentially happen per increment not per release but per increment. It helps you as a risk coping mechanism. This is by far one of the most important ones in the whole deck, it's problem solved, be a problem solver, solve the real problem, don't ask "mother may I sponsor? "may I" Don't do that, put on your thinking hat, solve the problem.

Offer solutions not problems. You go in to ask question, in many instance is the bad thing to do. The best thing to do is to solve the problem with the team and get stuff moving, move the project forward.

2.4 Systematic and Strategic: Think systematically and strategically to navigate complexity

This refers to systematic and strategic thinking. Think systematically and strategically to navigate complexity. You should ask the question; "What does the company want? What does the company need? What is the whole picture? What is going on right now?"

When they give you a question on the exam, you need to understand the question carefully to understand what is happening and you need to be able to see how everything is interwoven, how everything joins and you must weave the story together really quickly in under a minute. You've got to get the gist of the story and you've got to be able to navigate that complexity being given because that's what the exam is about; complexity, problem-solving, being strategic, answering the question; "What is the best strategic position based on this question?"

2.5 Change Management: Manage change & configuration with intentionality

Manage change and configuration with intentionality. Understand change is multi-fold. It is all about changes on at the project level and how project change could be a factor of organizational change and vice-versa. It could pertain to project documentation, plans and deliverables.

When we talk about configuration, we're specifically talking about artifacts and deliverables and version control of things like drawings and widgets and overall deliverable items.

Change management entails managing alterations to the project scope, timeline, or resources intentionally and effectively. Distinguishing between change management and configuration management is crucial to handle changes effectively on the project, including documents, artifacts, and deliverables. Proper version control is implemented for different project components to ensure consistent tracking and implementation.

Change can be scary, but as Agile teaches us, "Welcome changing requirements, even late in development. Agile processes harness change for the customer's competitive advantage." So, don't fear change; manage it. Keep an open mind, understand the difference between change management and configuration management, and remember, we're in control.

In Agile environments, change is welcomed more fluidly and dynamically. There is less red-tape and change is woven into the very practices of agile.

2.6 Inspect & Adapt: Continuously Inspect, adapt and integrate all levels & layers

Continuously Inspect, adapt and integrate all levels & layers.

Continuously inspecting, adapting, and integrating at all levels and layers is a fundamental principle in Agile and project management. It emphasizes the importance of ongoing evaluation, adjustment, and collaboration to ensure project success and deliver high-quality outcomes.

Inspecting refers to the practice of regularly assessing project progress, team performance, and the quality of deliverables. This involves conducting frequent reviews, evaluations, and assessments to identify any deviations from the desired outcomes or potential areas for improvement. By inspecting regularly, teams can catch issues early on, make necessary adjustments, and prevent potential problems from escalating.

Adapting is the process of making changes and adjustments based on the insights gained from the inspection phase. It involves being flexible and responsive to feedback, changing requirements, and evolving circumstances. Agile approaches encourage teams to embrace change and adapt their plans, processes, and deliverables accordingly. By adapting, teams can ensure that they stay aligned with the project goals and meet the evolving needs of stakeholders.

Integrating refers to the collaborative nature of Agile practices, where different levels and layers of the project are brought together and aligned. It involves integrating various components, processes, and perspectives to create a cohesive and unified project approach. This integration can occur within the team, across different teams or departments, or even with external stakeholders. By fostering integration, teams can leverage diverse expertise, ensure consistency, and enhance overall project performance.

Quality assurance and control play a vital role in the continuous inspection, adaptation, and integration process. Quality assurance involves implementing planned and systematic actions to provide confidence that the product or service will meet the specified quality requirements. It focuses on proactive measures, such as establishing quality standards, conducting reviews, and implementing quality management processes.

On the other hand, quality control is a reactive process that ensures the actual product or service being developed meets the specified quality requirements. It involves activities like testing, inspection, and verification to identify and resolve any defects or deviations. By implementing both quality assurance and control processes, teams can strive for excellence, deliver a high-quality product or service, and foster a culture of quality within the project team.

In summary, continuously inspecting, adapting, and integrating at all levels and layers ensures that projects stay on track, address issues promptly, and deliver high-quality outcomes. By embracing this principle, project teams can maintain agility, respond effectively to changes, and foster a culture of continuous improvement.

2.7 Problem Solve: Be a problem solver! Offer solutions and not problems!

As you embark on your journey to become a certified Project Management Professional (PMP), adopting a problem-solving mindset is paramount. A PMP® is not only responsible for managing projects but also for effectively addressing the myriad challenges that arise throughout the project lifecycle. By embracing the role of a problem solver, you position yourself as a valuable asset to your team and stakeholders.

A problem-solving mindset entails approaching obstacles and setbacks with a proactive and solutions-oriented mindset. Rather than dwelling on the problems at hand, you focus on identifying and implementing viable solutions. This approach not only facilitates progress but also demonstrates your ability to think critically, make informed decisions, and adapt to changing circumstances.

As a PMP, your role involves mitigating risks, resolving conflicts, and overcoming obstacles that could impede project success. By actively seeking solutions, you show your commitment to finding practical and effective ways to address challenges. Instead of merely pointing out problems, you take the initiative to propose actionable solutions that align with project goals and objectives.

Effective problem solving requires a systematic approach. Begin by thoroughly understanding the issue at hand, gathering relevant information, and analyzing the underlying causes. This analytical process enables you to identify the root causes of the problem and develop a comprehensive understanding of its impact on the project. By diving deep into the issue, you can avoid superficial or band-aid solutions and address the underlying factors that contribute to the problem.

Furthermore, effective problem solving involves collaboration and leveraging the diverse expertise within your team. Encourage open dialogue and create an environment where team members feel comfortable sharing their insights and perspectives. By fostering a culture of collective problem solving, you tap into a wealth of knowledge and creativity, increasing the likelihood of finding innovative and robust solutions.

In addition to proposing solutions, it is essential to consider their feasibility, potential risks, and anticipated outcomes.

Evaluate the impact of each proposed solution on project scope, timeline, and resources. This evaluation process allows you to weigh the pros and cons, assess trade-offs, and make informed decisions that align with the project's constraints and objectives.

Being a problem solver also involves continuous learning and improvement. Reflect on the effectiveness of the solutions implemented, gather feedback from stakeholders, and assess the outcomes. By evaluating the results, you can refine your problem-solving approach and enhance your ability to tackle future challenges.

Remember, as a PMP, your problem-solving skills are not limited to the exam or specific project scenarios. Cultivate a problem-solving mindset in your everyday interactions and professional endeavors. Embrace challenges as opportunities for growth and development, viewing them as a chance to showcase your ability to navigate complexities and deliver value.

2.8 Quality Assurance and The Iron Triangle: Proactively build in quality and manage the iron triangle

Proactively build in quality and manage the iron triangle. What am I really saying here? Remember in the Agile manifesto, we read; technical excellence and good design enhances your agility. Well technical excellence is part of quality. Managing the iron triangle is when you understand the importance of schedule, cost, and scope and their impact on quality.

Quality and the Iron Triangle are two fundamental aspects of project management that play a crucial role in project success. As you navigate your journey as a project manager, it is essential to proactively build in quality and effectively manage the Iron Triangle to deliver successful outcomes.

Quality, in the context of project management, refers to meeting or exceeding the expectations and requirements of stakeholders. It involves delivering a product, service, or result that is fit for purpose, free from defects, and meets the established standards and specifications. By prioritizing quality, you ensure that the project's deliverables are of the highest standard, satisfying customer needs and driving stakeholder satisfaction.

To proactively build in quality, start by establishing clear quality objectives and metrics at the beginning of the project. Define what quality means for your project, considering factors such as functionality, reliability, performance, and user experience. Collaborate with stakeholders to gain a comprehensive understanding of their expectations and incorporate those expectations into your quality planning.

Once quality objectives are defined, it is crucial to integrate quality into every phase of the project lifecycle. This includes incorporating quality assurance processes, conducting regular inspections and reviews, and implementing appropriate testing and validation techniques. By proactively identifying and addressing potential quality issues early on, you minimize the risk of costly rework, delays, and customer dissatisfaction.

Furthermore, managing the Iron Triangle is essential for effective project delivery. The Iron Triangle, also known as the Triple Constraint or Project Management Triangle, represents the interconnected relationship between scope, time, and cost. These three elements form the foundation of project management, and any changes to one of them inevitably impact the others.

To manage the Iron Triangle effectively, you must understand the interdependencies between scope, time, and cost. A change in scope, such as adding new features or requirements, may affect project timelines and increase costs. Conversely, a tight deadline or budget constraint may require adjustments to the project scope to ensure successful delivery. By recognizing these trade-offs, you can make informed decisions and strike the right balance to meet project objectives within the defined constraints.

In the world of Agile when we talk about the iron triangle, you should remember that it's flipped on its head, and instead of you having scope fixed like we do in traditional project management, we have flexible scope, but we have a fixed schedule and budget.

Your sprint in Scrum is a fixed timebox where you get stuff done that fits into it. Your budget is fixed by virtue of a team (that is pretty much fixed within certain bounds). We don't want to introduce new team members ad-nauseam. Instead, we want to make sure when we introduce new team members, it's because we absolutely have to and we don't want to go over the number of 3 – 10 max. And we're not doing this to increase our velocity, we're not doing this to get more and more done in a non-Agile way.

We always want to remember the five stages of team development, want to proceed with care. Don't just add team members to boost your velocity. That's a bad, bad way to do things. Instead, you want to do it mindfully, being aware of the five stages of team development and you want to make sure it's a good fit and does not exceed 10 (per the Scrum Guide) or close to that. The Agile Practice Guide mentions between 3 – 9 people. Team size and dynamics will ultimately affect project and product quality.

Quality assurance involves implementing a set of planned and systematic actions necessary to provide adequate confidence that a product or service will satisfy given requirements for quality. Quality control is a procedure intended to ensure that a product or service under development meets specified requirements. Together, these processes work to identify and resolve quality issues promptly and foster a culture of quality within the project team.

Have you ever heard the saying, "Measure twice, cut once"? In our world, this means having a rigorous quality assurance and control process in place. It's all about delivering the best possible result.

Regular monitoring and control mechanisms are vital for managing the Iron Triangle. Continuously track project progress, compare it against the baseline plan, and assess the impact of any changes on scope, time, and cost. Implement robust change management processes to evaluate and approve scope changes, ensuring they align with project objectives and constraints. By actively managing the Iron Triangle, you maintain transparency, make informed decisions, and keep stakeholders informed of any adjustments to project parameters.

The integration of quality and the Iron Triangle requires effective communication and collaboration with stakeholders. Engage with stakeholders to manage their expectations, clarify requirements, and obtain their input on quality objectives and trade-offs. By fostering open and transparent communication, you build trust and ensure alignment between project goals and stakeholder needs.

Lastly, continuous improvement is key to both quality and Iron

Triangle management. Regularly assess project performance, collect feedback from stakeholders, and identify areas for enhancement. Embrace lessons learned and leverage best practices to refine your quality processes and optimize the management of the Iron Triangle in future projects.

In summary, proactively building in quality and effectively managing the Iron Triangle are critical components of successful project management. By establishing clear quality objectives, integrating quality into every project phase, and managing the interdependencies of scope, time, and cost, you enhance project outcomes and stakeholder satisfaction. Through effective communication, collaboration, and continuous improvement, you can navigate the challenges and complexities of project management, delivering high-quality results within defined constraints.

2.9 Risk and Governance: Proactively manage risk, and governance

Proactively manage risk and governance. You got to remember governance is the framework within which authority is exercised.

The span and type of governance in every situation differs from project to project and opportunity to opportunity, but you need to understand and be comfortable with what governance is and the language around governance.

Also, consider risk management. How do you proactively manage risk instead of being reactive? Proactive risk management means planning how to manage risks, identifying the risks, performing the right analysis, be they qualitative or quantitative, and then planning the risk response. Also ensuring the response is proactive, doing it at the right time, using your influence as a project manager to make things happen, and ultimately monitoring risk. All that dialogue is helpful. For this reason, you do need to know at a minimum every process group and what you do there, every knowledge area what you do there, and each one of the 49 processes. You should know what you're doing, that is nonnegotiable.

Risk management is a proactive process that entails identifying, assessing, and controlling risks throughout the project lifecycle. Governance, on the other hand, provides the framework for decision-making and control within the project, ensuring proper authority and compliance. Both are interwoven to ensure that project goals are achieved while mitigating potential risks.

Here's where we need to be like a chess player, thinking several steps ahead and anticipating any obstacles that might come our way. We need to embrace the principle of "Build projects around motivated individuals. Give them the environment and support they need, and trust them to get the job done."

2.10 Manage All Areas: Logically plan and manage all knowledge areas

This summary simply states logically plan and manage all knowledge areas, and we're talking about integration all the way on down to stakeholders.

I know for the longest time I have said don't go crazy on the PMBOK® Guide, but just because I said don't go crazy on the book doesn't mean that you're going to go into the exam this week, and you can't explain integration or scope or cost! If you can't explain any of the knowledge areas, that's a problem my friends so even if you end up saying "Phill I can't read this book, it's so big" Look for my one-minute summaries on YouTube, watch them. There's no excuse. Look, I put out over 2000 videos for you. There's no excuse. If you do a search "Praizion risk", you'll come up with a plethora of videos. I've lost track of them but they're all out there for you. Tap into it, just look for the videos and watch.

You need to know the 10 knowledge areas, what they are, the five process groups, the whole story about them being done iteratively on some projects, and the 49 processes. And yes, I know it's not in the PMBOK® Guide 7th explicitly but PMI®

have that in the background in Standards Plus, which is linked to the 7th Edition. So if anyone is telling you, you don't need to know some technicalities, they're lying. And it's risky because people are still failing the PMP. We're not hearing about a lot of them but people are still failing in some numbers so you've got to be smart! Better your chances of success by really knowing what every knowledge area, process group, and process is!.

Manage all areas, understand all the areas, all the knowledge areas and how to effectively manage them. Look out for my series on 500 project management definitions for the PMP® Exam on YouTube.

2.11 Buy-In & Authorization: Seek authorization & buy-in when necessary

In the realm of project management, a powerful mindset mantra to embrace is to "Seek authorization and buy-in when necessary." This mindset underscores the importance of involving the team and stakeholders during critical decision-making processes, such as obtaining project authorization in a charter or beyond, and navigating trade-offs.

At various stages of your project, there will arise moments when seeking authorization becomes vital. This involves engaging key stakeholders, rallying their support, and ensuring alignment with project goals and objectives. By involving the right people, you harness their expertise and insights, fostering a sense of ownership and accountability within the team.

Moreover, project management often entails making tough choices and trade-offs. When faced with competing priorities or limited resources, securing buy-in becomes crucial. This involves persuading stakeholders, collaborating with them to find mutually beneficial solutions, and negotiating to reach consensus. By engaging in open discussions and considering different perspectives, you create an environment where trade-offs are understood and accepted.

Emphasizing persuasion, collaboration, and negotiation as keywords in this mindset underscores their significance. These skills enable effective communication, bridge gaps between diverse viewpoints, and build consensus. Through persuasive arguments, collaborative efforts, and skillful negotiation, you pave the way for informed decision-making and successful project outcomes.

Adopting the mindset of seeking authorization and buy-in when necessary empowers you to navigate project challenges with confidence, fostering a culture of collaboration and shared ownership. Remember, by harnessing the power of persuasion, collaboration, and negotiation, you lay the foundation for a resilient and successful project journey.

2.12 Closing: Close each stage, iteration or phase with a retrospective or lessons learned

One essential practice in project management is to "Close each stage, iteration, or phase with a retrospective or lessons learned." This practice, often emphasized in the Agile Practice Guide, holds significant importance and is regarded as one of the most critical meetings in the world of Agile or Scrum.

The retrospective serves as a powerful tool for fostering continuous improvement and embracing the concept of empiricism. It allows the team to reflect on their work, identify strengths and areas for improvement, and make informed decisions to enhance future performance. By dedicating time to analyze the process, gather insights, and capture lessons learned, the team can effectively iterate and refine their approach.

Within the realm of Scrum, the retrospective stands out as a cornerstone ceremony. It provides a dedicated space for the team to openly discuss successes, challenges, and opportunities for growth. Through this collaborative session, the team can work collectively to optimize their performance, drive innovation, and uncover areas where they can make tangible improvements.

The underlying principle behind the retrospective is to embrace empiricism, which is at the core of Agile approaches. By continually evaluating the process, experimenting, and adapting based on real-time feedback, the team can evolve and refine their practices. This iterative approach enables them to leverage their discoveries and insights, incorporate them back into the project pipeline, and continually enhance their effectiveness.

Ultimately, the goal of closing each stage, iteration, or phase with a retrospective is to foster a culture of continuous learning and improvement. By prioritizing this crucial meeting, teams can harness the power of reflection, collaboration, and knowledge sharing. They can identify patterns, address bottlenecks, celebrate achievements, and implement actionable steps to enhance their future performance.

Embracing the retrospective as a vital practice empowers teams to leverage their collective intelligence, refine their processes, and drive meaningful change. It reinforces the spirit of Agile approaches, where adaptability, transparency, and continuous improvement form the bedrock of project success.

The project review process involves identifying and recording lessons learned to improve future projects. These insights contribute to a continuous improvement process where strategies and processes are refined based on past experiences.

Just like how a surfer learns from each wave, we need to learn from each project. We need to capture and apply these lessons learned so that we don't make the same mistakes twice. It's all about building a culture of continuous improvement, where we get a little better each time. After all, as the Agile Manifesto says, "At regular intervals, the team reflects on- how to become more effective, then tunes and adjusts its behavior accordingly."

CHAPTER THREE: BUSINESS PERFORMANCE DOMAIN

Let's move into the Business Domain. Here's a final one business, when you're tackling the world of business on the exam, be prepared to hear words such as outcomes, value, benefits, revenues, advantage, demand, cost of delay, that's how you should be thinking even if you don't have that exact one on the exam. Cost of delay, revenue, leakage, opportunity erosion, competition compliance, organization change, strategy, and business, all of these are topics and ideas you should be familiar with.

A business mindset is essential for project managers who want to be successful. It means understanding the business needs of the project and how the project can deliver value to the organization. It also means being able to communicate effectively with stakeholders and manage the project in a way that aligns with the organization's goals.

Business acumen is like having a sixth sense. It's the ability to make sharp, quick decisions that drive your business forward. As the Oracle of Omaha, Warren Buffet, advised, "Risk comes from not knowing what you're doing." Business acumen is about knowing.

Business acumen is the ability to understand and apply business concepts and principles. This includes the ability to understand the organization's business model, the ability to make sound business decisions, and the ability to manage the organization's resources effectively.

Developing strong business acumen is essential for project leaders to make informed decisions and understand the broader implications of their projects. John Maxwell's quote, "Leaders must be close enough to relate to others, but far enough ahead to motivate them," emphasizes the importance of striking a balance between operational insights and visionary thinking.

"Business acumen is the ability to see the big picture and make decisions that are in the best interests of the organization." – Unknown

A key topic to understand in this section is STRATEGY! Your company's strategy and culture are like its DNA - they're what make you unique. They're the roadmap that guides all your decisions. As Peter Drucker said, "Culture eats strategy for breakfast."

Organizational strategy is the plan that the organization uses to achieve its goals. This plan includes the organization's goals, the strategies that the organization will use to achieve those goals, and the resources that the organization will need to achieve those goals.

"Strategy is the art of making choices." - Michael Porter Organizational culture is the set of values and beliefs that guide the organization's behavior. This culture influences how the organization operates, how it makes decisions, and how it interacts with its stakeholders.

Understanding the organization's strategy and culture is vital for project leaders to navigate internal dynamics successfully.

3.1 Environment: Observe & respond to the external and internal environment

One crucial aspect to consider in project management is the "Environment." As a boundary spanner, someone who bridges the gap between your company and other companies, as well as between your company and the business environment, it is vital to observe and respond accordingly.

This is all about understanding the playground we're playing in. It's like a surfer knowing the tides or a pilot understanding the weather. You have to know the business environment to navigate it effectively. As Sun Tzu said in The Art of War, "If you know the enemy and know yourself, you need not fear the result of a hundred battles."

As a product owner, you play a pivotal role in adjusting your product backlog based on the ever-changing environment. In the 7th edition, emphasis is placed on understanding both the internal and external project environment. This includes factors within the project itself and external influences. The concept of Enterprise Environmental Factors (EEFs) and Organizational Process Assets (OPAs) is now integrated into this internal/external framework. Understanding the nuances of a product owner's responsibilities is crucial as the exam evaluates not only project management but also the interaction of product owners, sponsors, and the project team.

The environment holds significant weight in project management. It encompasses a broad spectrum of factors that can impact project outcomes. By diligently observing and staying attuned to the environment, you can proactively respond to emerging challenges, capitalize on opportunities, and make informed decisions.

Being aware of the internal and external dynamics enables you to navigate complexities, anticipate potential risks, and adapt your product backlog to align with the evolving circumstances. Your role as a boundary spanner necessitates understanding the broader context in which your project operates, including market trends, regulatory changes, technological advancements, and organizational influences.

Recognizing the significance of the environment prepares you for the diverse interactions you may encounter as a product owner. The exam will assess your ability to effectively collaborate with project managers, product owners, sponsors, and the project team. It emphasizes the need for a comprehensive understanding of the environment's impact on project delivery and success.

The environment plays a pivotal role in project management, particularly for product owners. By embracing your role as a boundary spanner and actively observing and responding to the internal and external environment, you can optimize your decision-making, enhance your product backlog, and contribute to project success. Understanding the multifaceted nature of your responsibilities as a product owner is crucial for the exam and your professional practice.

Business environment awareness is the ability to understand the factors that impact the organization's business. This includes factors such as the economic climate, the competitive landscape, and the regulatory environment.

"The business environment is constantly changing, and we need to be able to adapt to those changes." - Michael Porter

Being aware of the broader business environment is crucial for project leaders to make informed decisions. Bill Gates once stated, "We always overestimate the change that will occur in the next two years and underestimate the change that will occur in the next ten. Don't let yourself be lulled into inaction." A real-world example of business environment awareness can be seen in the case of the COVID-19 pandemic. The COVID-19 pandemic has had a significant impact on many businesses, and it has forced businesses to adapt their operations in order to survive.

Analyzing the market and industry landscape is crucial to identify opportunities and stay ahead of the competition. Tom Peters once stated, "If a window of opportunity appears, don't pull down the shade." This quote emphasizes the importance of seizing opportunities that arise from astute market and industry analysis.

Market and industry analysis is like a treasure hunt. It's about exploring the market, understanding the competition, and identifying opportunities. Henry Ford, the founder of the Ford Motor Company, once said, "The competitor to be feared is one who never bothers about you at all, but goes on making his own business better all the time."

Market and industry analysis is 0the process of understanding the market and industry in which the organization operates.

This includes understanding the needs of the customers, the competitive landscape, and the trends that are shaping the market.

"The market is constantly changing, and we need to be able to understand those changes." - Michael Porter

A real-world example of market and industry analysis can be seen in the case of the rise of e-commerce. The rise of e-commerce has had a significant impact on the retail industry, and it has forced retailers to adapt their strategies in order to compete.

Real-world dialog: In a strategy meeting, the project leader presents the findings of a comprehensive market and industry analysis. One team member asks, "How can we leverage these insights to gain a competitive edge?" The project leader responds, "Let's brainstorm ideas on how we can adapt our project approach to capitalize on the emerging market trends and consumer preferences."

3.2 Outcomes: Focus on outcomes, value and benefits over output

As a project manager, one important mindset to adopt is a focus on outcomes rather than mere output. While output refers to the tangible deliverables and tasks completed throughout the project, outcomes encompass the broader goals, value, and benefits that the project aims to achieve.

By shifting your perspective towards outcomes, you prioritize the ultimate impact and value that the project will create for stakeholders and the organization as a whole. It involves understanding and aligning with the strategic objectives and desired benefits of the project.

Instead of solely focusing on completing tasks and producing deliverables, you actively seek to understand the underlying purpose and intended outcomes. This involves engaging with stakeholders to clarify their expectations and desired results. By understanding their needs and aspirations, you can better shape the project's direction and outcomes. Throughout the project lifecycle, you continuously assess whether the project is on track to deliver the intended outcomes. This includes monitoring key performance indicators (KPIs) and milestones that directly contribute to the realization of benefits. Regularly evaluating progress allows you to make necessary adjustments and course corrections to ensure the desired outcomes remain in focus.

Value realization becomes a guiding principle in decision-making. It involves considering the cost-benefit analysis, prioritizing activities that yield the most significant value, and optimizing resource allocation to maximize outcomes. This may involve making trade-offs, reevaluating priorities, or even challenging existing assumptions to ensure that the project is aligned with its intended purpose.

By emphasizing outcomes, you foster a results-driven mindset within the project team. You encourage collaboration, innovation, and problem-solving geared towards achieving the desired benefits. Team members are motivated to think beyond the immediate tasks and consider the broader impact of their efforts. Communicating the focus on outcomes to stakeholders is also crucial. By highlighting the value and benefits that the project aims to deliver, you gain their support and buy-in. Clear and transparent communication about the project's progress, milestones achieved, and anticipated outcomes helps maintain stakeholder engagement and alignment.

Ultimately, by prioritizing outcomes, value, and benefits, you ensure that the project's purpose remains at the forefront. This mindset encourages a holistic and strategic approach to project management, resulting in greater satisfaction for stakeholders, enhanced organizational value, and a higher likelihood of achieving long-term success.

During the exam, expect the term "outcomes" to be emphasized. Focus on outcomes, not just deliverables or outputs. While deliverables are important, your primary focus should be on the ultimate outcome. Consider value, benefits, and desired outcomes over mere output. For instance, imagine a software project where the goal is to have people use and benefit from the software. Merely delivering the software as an output may not lead to the desired outcome if people continue to use alternative systems. In such cases, the real benefits and value of the software aren't realized. To ensure the desired outcome, consider including activities like a cutover period in your backlog. These additional measures will help ensure the true outcome is achieved.

3.3 Organizational Change: Set the stage for organizational change and build alliances.

Change is a huge part of the PMP® Exam! Let's expand on the mantra.

1. Understand the need for change: The first step in setting the stage for organizational change is recognizing and articulating the need for it. The trigger could be external, such as market shifts, changes in technology, or new regulations, or it might be internal, like a strategic shift, changes in leadership, or a desire for improved efficiency. Thoroughly understanding and articulating this need is the first step in convincing others to embark on this journey.

2. Develop a clear vision and strategy: Once you've identified the need for change, develop a clear vision of what the organization will look like after the change. This vision should be compelling and easy to understand. Additionally, a strategy outlining how you will achieve this vision should be established. This includes identifying key milestones, resources needed, potential risks and mitigation strategies, and a timeline.

3. Build alliances: Change is rarely achieved in isolation. As a project manager, you'll need to build alliances

across the organization to support and advocate for the change. This includes key stakeholders, senior leaders, other managers, and even individual contributors who can influence others. Engage these allies early, solicit their feedback, and involve them in planning and executing the change.

4. Communicate consistently and transparently: Throughout the change process, clear and consistent communication is vital. Share the vision, strategy, and progress regularly with your allies and the wider organization. Address any concerns, respond to feedback, and keep everyone updated on milestones and setbacks. Transparency builds trust, which is critical for successful change.

5. Manage resistance: Resistance to change is natural and should be expected. As a project manager, it's your role to understand the source of this resistance and address it. This could be through more effective communication, additional training, or addressing concerns directly.

6. Celebrate success: Even small wins along the way should be celebrated. This helps to build momentum

and shows the organization that progress is being made.

7. Embed the change: Finally, for change to be lasting, it must become part of the organization's DNA. This could involve updating policies and procedures, changes to job roles, or continuous training and support. Embedding the change ensures that it sticks even after the project is completed.

In conclusion, successful organizational change is a complex process that requires careful planning, building alliances, consistent communication, managing resistance, celebrating success, and embedding the change. As a project manager, understanding and applying these steps can greatly increase the likelihood of your change initiatives succeeding.

3.4 Project Impact: Assess the project's impact on the organization and navigate accordingly.

As a project manager, it is crucial to recognize and assess the impact of the organization on your project and navigate accordingly. Organizations can have a significant influence on project outcomes, and understanding how to navigate through organizational dynamics is essential for success.

When the organization exerts a substantial impact on the project, it means that various factors such as organizational structure, culture, policies, and stakeholders can significantly shape the project's environment. As a project manager, you need to be proactive in identifying and understanding these influences to effectively navigate through them.

One aspect of managing organization impact is recognizing the organizational structure and hierarchy. Understanding reporting lines, decision-making processes, and communication channels helps you engage with the right stakeholders, seek necessary approvals, and effectively communicate project updates and progress. This knowledge enables you to navigate through the organization's structure to obtain the required support and resources for your project.

Additionally, being aware of the organizational culture is crucial. Each organization has its unique set of values, norms, and behaviors that shape how projects are approached and managed. By understanding the prevailing culture, you can adapt your project management approach, communication style, and stakeholder engagement strategies to align with the organization's cultural expectations. This adaptability fosters better collaboration, reduces resistance, and enhances project success.

Policies and procedures within the organization can also impact your project. Familiarize yourself with relevant policies, guidelines, and frameworks that govern project execution. This knowledge helps you ensure compliance, mitigate risks, and align your project with the organization's strategic objectives. It is important to work within the established policies while also advocating for necessary adjustments or exceptions when they align with project goals.

Navigating through organizational dynamics requires effective stakeholder management. Identify key stakeholders, understand their roles, interests, and influence, and proactively engage with them throughout the project.

Building relationships, managing expectations, and addressing concerns or conflicts are essential for successfully managing the impact of the organization on your project.

As a project manager, your ability to navigate the organization's impact is a critical skill. It involves assessing the organizational landscape, adapting your approach, and leveraging relationships and influence to secure the necessary support for your project. By understanding and working within the organizational context, you can effectively manage challenges, mitigate risks, and optimize project outcomes.

Assessing Project Impact: As a project manager, understanding the project's potential impact on your organization is crucial. This means not just looking at the immediate project deliverables, but also considering the broader implications. What strategic objectives does the project align with? How will it affect various departments or teams? Will it cause any significant shifts in workflow or operations?

The project impact can be assessed in various ways, such as conducting an Impact Analysis, which involves identifying all the possible effects and changes that a project might bring about within the organization. This could include positive impacts such as improved processes, increased revenue, or enhanced customer satisfaction. But you also need to consider potential negative impacts, like disruption of day-to-day operations, temporary dip in productivity, or possible resistance to change from employees.

Navigating Accordingly: Once you've assessed the project's impact, the next step is to navigate the project execution based on this understanding.

Developing a Communication Plan: This is crucial to keep everyone involved informed about what the project will entail, how it will affect them, and what their roles will be. Regular, clear, and transparent communication helps to manage expectations, address concerns, and increase buy-in.

Mitigation and Contingency Planning: If the assessment reveals potential negative impacts, you'll need to develop strategies to mitigate these risks. This could involve scheduling certain tasks to minimize disruption, providing additional training or resources, or developing contingency plans in case issues arise.

Change Management: Any project bringing about significant change should have a thorough change management strategy. This involves helping the organization and its people adjust to the new changes smoothly, which might require activities like coaching, training sessions, or even team-building exercises.

Monitoring and Adjusting: Once the project is underway, it's important to continually monitor its impact and make any necessary adjustments. This could be in terms of scope, resources, timeline, or even the project's strategic alignment.

Remember, assessing the project's impact and navigating accordingly is not a one-off task at the beginning of the project. It's a continuous process that should be revisited throughout the project lifecycle to ensure the project remains aligned with the organization's goals and adapts to any changes that might occur along the way.

3.5 Organization Impact: Assess the organization's impact on the project & navigate accordingly

When assessing the organization's impact, it involves evaluating various factors that can shape the project environment. These factors include the organization's structure, culture, policies, resources, and stakeholders. By thoroughly assessing these aspects, you gain insights into how they can either support or hinder project progress. Let's expand this mantra for deeper understanding:

1. Assessing Organization Impact: As the project manager, it's your job to understand how the organization's culture, structures, and resources could impact your project. For example, the organization's culture might affect the speed of decision-making, the level of innovation, and team collaboration. Bureaucratic structures might slow down approvals or affect resource allocation, while a flat organizational structure could speed things up but might also blur lines of accountability. You should also consider the organization's resources and capabilities, which can have a significant impact on your project. This could include financial resources, human resources,

technology, and infrastructure. Are the necessary resources available and accessible? Does the organization have the capabilities to achieve the project goals? In addition, understanding the organization's strategic goals and priorities is important. Is your project a top priority for the organization? If not, it might be more difficult to secure resources or stakeholder support.

2. Navigating Accordingly: Once you've assessed the organization's impact, the next step is to navigate the project based on this understanding.

3. Stakeholder Management: The organization is full of stakeholders, each with their own interests, priorities, and concerns. Effective stakeholder management involves identifying key stakeholders, understanding their needs and concerns, and managing their expectations throughout the project.

4. Resource Management: If resources are a concern, you might need to get creative with your resource management. This could involve prioritizing tasks based on resource availability, making a case for additional resources, or finding ways to do more with less.

5. Navigating Organizational Structures: Understanding the organization's structures and decision-making processes can help you navigate through any bureaucratic hurdles. It might also help you identify potential allies or blockers who could impact your project.

6. Aligning with Organizational Goals: Your project is more likely to be successful if it's clearly aligned with the organization's strategic goals. This could involve highlighting how your project contributes to these goals, or it might involve adjusting your project goals to better align with the organization's priorities.

7. Change Management: If your project involves significant change, understanding the organization's readiness and capacity for change is crucial. This might involve conducting a change readiness assessment, or it might involve additional activities to prepare the organization for change.

Overall, understanding the organization's impact on the project and navigating accordingly can help you anticipate challenges, manage resources effectively, and keep your project aligned with the organization's strategic goals.

3.6 Proactively ensure management of benefits & their realization

Benefits realization is the driving force behind every successful project. As Steve Jobs once famously said, "It's not about ideas, it's about making ideas happen." This sentiment captures the essence of benefits realization. It's about turning ideas into tangible results that bring value to the organization. When you go fishing, you're not just hoping to catch a fish, you're aiming for a satisfying meal. Likewise, in business, we always have our eyes on the benefits - the real, tangible returns we want from our projects. As Benjamin Franklin once said, "An investment in knowledge pays the best interest."

Benefits realization is the process of ensuring that the project delivers the intended benefits to the organization. This involves identifying the benefits, measuring the benefits, and tracking the benefits to ensure that they are realized.

"The project is not over when the product is delivered. The project is over when the benefits are realized." - Peter Drucker

A real-world example of benefits realization can be seen in the case of the iPhone. The iPhone was a major success for Apple, but it wasn't just the product that was successful. The iPhone also delivered a number of benefits to Apple, such as increased sales, increased market share, and increased brand awareness.

3.7 Value Swap: Swap out backlog items with work of comparable value

In project management, especially when applying agile approaches, you often have a backlog - a list of tasks or features that need to be addressed in the project. But sometimes, due to changes in circumstances, requirements, or priorities, you might need to swap out some backlog items.

When you do a value-swap, you're exchanging one backlog item for another that brings comparable value to the project. It's essential to remember here that value doesn't necessarily mean effort or time; it's more about the return on investment or the impact on the end product.

When to Apply Value-Swap: There are several situations when a value-swap might be appropriate:

- Changing Priorities: If the project's priorities change, it might be necessary to bring in new tasks that better align with these priorities, even if that means swapping out existing backlog items.

- Feedback and Learning: If you receive new information or feedback that changes your understanding of what's valuable, a value-swap might be in order.

- Risk Management: If you identify a new risk that needs to be addressed immediately, swapping in a task to mitigate this risk might be the right move.

How to Navigate Value-Swap: Navigating a value-swap effectively requires a solid understanding of the project, good communication, and a willingness to adapt:

- Understand the Value: To swap out backlog items, you first need to understand the value each item brings to the project. This might involve discussing with stakeholders, using data or metrics, or drawing on your own experience and expertise.

- Communicate Effectively: Any changes to the backlog should be communicated clearly to the project team and relevant stakeholders. Explain why the swap is happening, what the new task is, and why it's of comparable value.

- Stay Adaptable: The value of different backlog items might change as the project progresses. Stay open to this possibility, and be ready to make further swaps if needed.

Remember, the goal of value-swapping isn't to increase the workload or squeeze in more tasks; it's to ensure the project's work remains valuable and aligned with the project's goals and priorities. This approach fosters a more responsive, adaptable, and value-focused project management style.

3.8 Value Delivery: Strategically plan the value delivery system

Understanding Value Delivery System (VDS): VDS is a term coined by PMI in the PMBOK® Guide Seventh Edition. A VDS encompasses everything from individual projects to large-scale operations. It's a strategic way of organizing projects, programs, portfolios, and operations to maximize value and benefits. It represents an integration of all these elements working seamlessly together to deliver value to stakeholders and the organization.

Strategically Planning the VDS:

- Align with Organizational Strategy: The first step in planning the VDS is ensuring alignment with the organization's overall strategy and goals. This alignment should permeate all levels of the system, from the portfolio level down to individual projects.

- Prioritize Based on Value: Not all projects or programs offer the same value. Prioritize those that promise to deliver the highest value aligned with strategic objectives. This involves a clear understanding of what 'value' means within your organization – it could be financial gain, customer satisfaction, risk mitigation, innovation, etc.

- Resource Allocation: Strategically allocate resources based on value prioritization. This doesn't mean all resources go to the highest-value projects or programs; instead, it's about finding a balance to maintain progress across the board while maximizing overall value delivery.

- Integration and Coordination: Ensure seamless integration and coordination between projects, programs, portfolios, and operations. Avoid silos and foster collaboration and information sharing for efficient value delivery.

Navigating the VDS: Once your strategic plan is in place, it's time to navigate through the VDS.

- Ongoing Monitoring and Evaluation: Track performance, deliverables, and outcomes regularly. Use key performance indicators (KPIs) that reflect the value being delivered. This helps in identifying any deviations and enables timely interventions.

- Adapt and Adjust: Be ready to adapt and adjust as things change – whether it's internal factors like resource availability or external factors like market changes. The VDS should be flexible enough to accommodate such changes without compromising on value delivery.

- Stakeholder Engagement: Keep stakeholders engaged and informed about progress and any changes. Their continued buy-in and support are vital for the VDS's success.

- Feedback and Continuous Improvement: The VDS isn't a set-it-and-forget-it system. Regularly collect feedback, learn from successes and failures, and look for ways to improve value delivery.

Value delivery is the process of ensuring that the project delivers value to the customer. This involves understanding the customer's needs, designing a product or service that meets those needs, and delivering the product or service in a way that meets the customer's expectations.

Think of value delivery like a pizza delivery. You want your pizza hot, tasty, and on time. Our customers expect the same from us - delivering valuable products and services that meet their needs. Jeff Bezos, the founder of Amazon, once said, "We see our customers as invited guests to a party, and we are the hosts."

Value delivery is the key to maximizing return on investment and ensuring that projects align with business objectives. Bill Gates once said, "Your most unhappy customers are your greatest source of learning." This quote underscores the importance of understanding customer needs and delivering value that exceeds their expectations.

"The customer is the only one who can define value." - Peter Drucker

A real-world example of value delivery can be seen in the case of Amazon. Amazon has been very successful in delivering value to its customers by offering a wide variety of products, competitive prices, and convenient delivery options.

Business Strategy Alignment: This is an associated topic with the VDS. Business strategy alignment is the process of ensuring that a project is aligned with the organization's overall business strategy. This involves understanding the organization's goals, identifying the projects that will help the organization achieve those goals, and managing the projects in a way that supports the organization's strategy.

"The project is the means, not the end." - Peter Drucker

Aligning project objectives with the broader business strategy is crucial for long-term success. Tom Peters, a renowned management expert, once stated, "Strategy is about making choices, trade-offs; it's about deliberately choosing to be different." This quote emphasizes the need for project leaders to align their initiatives with the overall strategic direction of the organization.

This is like aligning the wheels of a car. If they're not in sync, the ride can be bumpy and inefficient. Similarly, our projects need to align with the business strategy for a smooth ride to success. As the great Michael Porter said, "Strategy is about making choices, trade-offs; it's about deliberately choosing to be different."

A real-world example of business strategy alignment can be seen in the case of the Ford Mustang. The Ford Mustang was a major success for Ford because it was aligned with the company's overall business strategy. The Mustang was a sporty car that appealed to a younger demographic, which was a key market segment for Ford at the time.

Sarah, a project leader, recounts an experience where alignment with business strategy led to project success. "We were tasked with developing a new digital marketing campaign. By thoroughly understanding our company's brand positioning and target audience, we were able to create a campaign that aligned perfectly with our overall marketing strategy. As a result, we saw a significant increase in brand awareness and customer engagement."

3.9 Compliance: Proactively manage compliance

Compliance is a critical aspect of project management that should not be overlooked. As a project manager, it is essential to proactively manage compliance to ensure that your project operates within the established legal, regulatory, and organizational frameworks. The PMP® exam places a strong emphasis on assessing your understanding and awareness of compliance and your ability to navigate it effectively.

Being in compliance means adhering to applicable laws, regulations, industry standards, and internal policies. Failure to comply can have serious consequences, including legal and financial implications, reputational damage, and project delays or disruptions. Therefore, project managers need to have a comprehensive understanding of the compliance requirements relevant to their projects.

To proactively manage compliance, you must stay informed about the legal and regulatory landscape related to your project. This involves conducting thorough research, consulting with subject matter experts, and staying up-to-date with any changes or updates that may impact your project's compliance obligations.

Navigating compliance also requires implementing robust control mechanisms and processes. This includes establishing clear documentation practices, conducting regular audits and assessments, and implementing corrective actions when non-compliance is identified. It's important to prioritize compliance-related activities, allocate resources accordingly, and communicate effectively with stakeholders to ensure a shared understanding and commitment to compliance.

During the PMP® exam, expect questions that evaluate your knowledge of compliance frameworks, risk assessment and mitigation strategies, documentation requirements, and stakeholder management in the context of compliance. Demonstrating your ability to proactively manage compliance and mitigate compliance-related risks will showcase your competence as a project manager.

By demonstrating a strong understanding of compliance, you will not only excel in the PMP® exam but also establish yourself as a responsible and trusted project manager in real-world scenarios. Remember, compliance is not just a box to check off—it is a crucial element for successful project delivery and stakeholder satisfaction.

3.10 Sustainable Community: Harness COPs, PMOs & VDOs for the firm's strategic goals

Creating a sustainable community within an organization involves harnessing the collective power of Communities of Practice (COPs), Project Management Offices (PMOs), and Virtual Delivery Organizations (VDOs) to align with and support the firm's strategic goals. These entities play crucial roles in fostering collaboration, knowledge sharing, and driving organizational success. Let's break this down:

1. **Understanding the Sustainable Community:**

A sustainable community, in this context, refers to a network of individuals and groups within an organization that supports its strategic objectives over the long term. The key players are Communities of Practice (COPs), Project Management Offices (PMOs), and Value Delivery Offices (VDOs).

- **Communities of Practice (COPs):** These are groups of people who share a common interest or expertise and come together to share knowledge, solve problems, and innovate. They span across the organization, cutting through hierarchical and departmental boundaries, and foster a culture of continuous learning and collaboration.

- **Project Management Offices (PMOs):** These are dedicated entities within an organization that oversee project management. They provide support, standards, methodologies, and oversight to ensure projects are managed consistently and effectively.

- **Value Delivery Offices (VDOs):** These entities focus on ensuring that projects, programs, and portfolios deliver their intended value. They coordinate and integrate these efforts to align with the organization's strategic goals.

2. **Harnessing COPs, PMOs, and VDOs for Strategic Goals:**

  - **Align with Strategy:** Align the focus and activities of COPs, PMOs, and VDOs with the firm's strategic goals. This may involve aligning the COPs around key strategic areas, ensuring that PMO methodologies and standards reflect strategic priorities, and aligning VDOs with strategic objectives.

  - **Promote Collaboration:** Encourage collaboration between COPs, PMOs, and VDOs. This might involve joint meetings, cross-functional projects, or shared tools and platforms. Collaboration can lead to more integrated and effective strategic execution.

- **Leverage Expertise:** Utilize the expertise within COPs to enhance project execution and value delivery. This might involve drawing on COP members for training, consultation, or innovation.

- **Emphasize Value Delivery:** Through the VDOs, maintain a focus on delivering value that aligns with strategic goals. This might involve regularly evaluating and adjusting projects, programs, and portfolios to ensure they are delivering their intended value.

- **Maintain Oversight and Consistency:** Use PMOs to maintain oversight and consistency in project management. This can help ensure that projects are executed efficiently and effectively, contributing to strategic goals.

3. **Nurturing the Sustainable Community:**

- **Provide Support:** Provide resources and support to COPs, PMOs, and VDOs. This might involve training, tools, or dedicated time and space for COP activities.

- **Promote Learning and Growth:** Encourage continuous learning and growth within COPs, PMOs, and VDOs. This might involve regular training, knowledge sharing sessions, or learning from project post-mortems.

- **Recognize and Reward:** Recognize and reward the contributions of COPs, PMOs, and VDOs. This can motivate members and reinforce their importance to the firm's strategic goals.

By harnessing COPs, PMOs, and VDOs, firms can create a sustainable community that supports their strategic goals, promotes a culture of collaboration and continuous learning, and ultimately enhances their ability to deliver value consistently and effectively.

3.11 Lean Thinking: Think and be lean to eliminate waste at all levels of the value delivery system

Embracing the mantra of Lean Thinking entails adopting a mindset focused on eliminating waste at all levels of the value delivery system. Lean principles, derived from the Toyota Production System, emphasize maximizing value while minimizing waste, thereby enhancing efficiency, productivity, and customer satisfaction.

1. **Understanding Lean Thinking:** Lean Thinking is a business methodology that emerged from the Toyota Production System (TPS). It's focused on delivering maximum value to the customer while minimizing waste. In this context, 'waste' refers to any resource expenditure that doesn't contribute to creating value for the customer.

2. **Thinking Lean:** This refers to the mindset you need to adopt to effectively implement lean principles. It involves looking critically at your business processes and identifying areas where value is not being added. In essence, you are constantly asking, "How can this be done more efficiently? Where is the waste, and how can it be eliminated?"

Thinking lean involves developing a critical eye to examine each step of a process and question its value contribution. It requires challenging the status quo, seeking innovative solutions, and continuously improving processes to enhance efficiency and eliminate non-value-added activities. This mindset encourages individuals and teams to adopt a problem-solving approach, looking for ways to streamline operations, eliminate redundancies, and enhance value creation.

3. **Being Lean:** This is the practical application of lean thinking. It involves making changes in your business processes to eliminate waste and increase value. This could mean removing unnecessary steps in a process, reducing resource use, improving coordination and communication, or any other changes that result in more efficient and effective operations.

4. **Eliminating Waste at All Levels:** Lean thinking should be applied across the entire value delivery system, not just at specific points or in specific processes. This includes everything from initial product design or service planning stages to delivery to the customer.

5. **Value Delivery System:** This refers to all the components that contribute to the creation and delivery of a product or service (projects, programs, portfolios and operations). By applying lean thinking, the goal is to ensure that every step in this system is optimized to deliver maximum value.

Eliminating waste at all levels of the value delivery system requires a holistic approach. It involves not only streamlining internal processes but also optimizing supplier relationships and customer interactions. Lean organizations engage in value stream mapping to visualize and analyze end-to-end processes, identifying areas for improvement and implementing lean solutions. This collaborative approach allows organizations to identify opportunities for waste reduction, enhance collaboration, and deliver greater value to customers.

To think and be lean means to cultivate a deep understanding of the different types of waste that can occur throughout the value delivery system. These wastes, known as the "Seven

Wastes" in Lean Thinking, include overproduction, waiting, transportation, unnecessary inventory, motion, over-processing, and defects. By actively identifying and eliminating these wastes, organizations can optimize their processes, reduce costs, and improve overall performance.

Being lean goes beyond individual actions and extends to the entire organization. It involves creating a culture that values waste elimination and empowers employees at all levels to contribute to process improvements. Organizations can establish lean practices by implementing visual management systems, fostering open communication, and providing training on Lean methodologies. By embedding lean thinking into the organizational DNA, continuous improvement becomes a shared responsibility, driving efficiency, and enabling sustainable growth.

Furthermore, lean thinking extends beyond operational processes and can be applied to decision-making, resource allocation, and strategic planning. Organizations can prioritize initiatives that align with customer needs, eliminate non-essential activities, and allocate resources effectively to maximize value creation. By thinking lean at all levels of the organization, companies can become agile, responsive, and competitive in dynamic business environments.

3.12 Gating: Use toll gates, stage gates, kill-points, and phase-end reviews to deliver only value

Gating is a key concept in project management that emphasizes the importance of using toll gates, stage gates, kill-points, and phase-end reviews to ensure that projects deliver only value at each stage of the process. These gating mechanisms act as checkpoints or decision points that help assess the progress, viability, and alignment of a project with organizational goals and objectives.

1. **Understanding Gating:** Gating, in a project management context, is a method for controlling the flow of a project's progress. It involves setting up checkpoints, or 'gates', at different stages in the project lifecycle. These gates serve as decision points where project leaders review the progress and decide whether to proceed, make changes, or cancel the project.

2. **Toll Gates and Stage Gates:** Toll gates, also known as stage gates, are points in the project where key stakeholders review the project's status and decide whether it should proceed to the next phase. The project is assessed on various factors, such as performance against the planned schedule, budget, quality, risk, and benefits realization. Only after clearing

the gate does the project move forward.

3. **Kill Points:** Kill points are similar to toll gates but with a specific focus on determining whether a project should be discontinued. They serve as decision points where, if the project is not delivering the expected value or if the risks are too high, it can be 'killed' or terminated to avoid further resource expenditure.

4. **Phase-End Reviews:** These are formal reviews conducted at the end of each project phase to evaluate the project's performance and align the next steps. They involve detailed assessments of the work done in the current phase and planning for the upcoming phase. Phase-end reviews help ensure that the project remains aligned with its objectives and that only value-adding activities are carried forward.

5. **Delivering Only Value:** The primary aim of gating is to ensure that the project delivers maximum value. By having these decision points, the organization can continually assess whether the project is on track to deliver the anticipated benefits. If it's not, corrections can be made, or in the case of a kill point, the project can be discontinued. This way, the organization ensures that resources are used efficiently and

effectively, focusing on activities that bring the most significant value.

Overall, the gating methodology promotes discipline in project management, enabling a controlled, phased approach to project progression. It provides a structured process for decision-making, ensuring that projects are continually evaluated and aligned with strategic goals. This approach helps organizations manage risks, control costs, and, ultimately, deliver more value.

PART 2: Agile Thinking & Practices for Exam Success

CHAPTER FOUR: AGILE DEEP THINKING

Let's first of all, examine Agile deep thinking!

When I say think Agile, I mean espouse the Agile Manifesto details inside out. You need to be thinking about people first, communicating as often as possible.

Remember face-to-face communication is preferred, you also need to have a positive mental attitude.

The questions will pose challenges in that, you could choose a negative mindset, one of punishment power or you could choose one of mentoring and coaching. If a team member is not performing up to par, what are you going to do? Punish them? or would you train and coach them? Would you help them to understand better ways of doing things?

Acceptable and Professional Servant-Leader Behavior

1. Be nice to people. Punishment is rarely a good course of action. Sometimes that people need to be given a slap on the wrist for poor choices and bad behavior in an organization yes, but that is rarely PMI's focus on these questions.

2. Cross-functional teams are important. It's important to see cross-functionality within the team where people are willing and able to jump in to help each other. That is the entire idea when we talk about cross-functional teams; T-shaped skills, broken comb, paint-drip. Those are words you should know.

3. Work with the customer relentlessly, have a collaborative mindset as opposed to one of negotiation, understand that change is good. Kaizen – change for the better. When your customer needs a change, give them the change to the extent possible. Empower the team.

4. Collaborate! To think and be Agile, you need to have a collaborative mindset when it comes to working with

teams to manage conflict. When it comes to leading a team, think servant leadership, think situational leadership, not a one-size-fits-all for everyone, but everyone on that team is a unique individual. If you are working with a team, work with them with the Hersey-Blanchard Model at the back of your mind. As a leader in an organization, you should support the team's performance. Supporting their performance means you are helping them grow as you understand their performance and give them feedback in a friendly non-threatening manner. The exam will test you on Agile thinking across people, process, and business domains.

5. The understanding of building a team, equipping the team, appraising their skills is important. When it comes to the team being stuck, you should remember your definite chief aim as a Scrum Master or servant leader is to continually reassess those impediments, obstacles, and blockers. Get them out of the way, prioritize them, big rocks first, smaller pebbles, sand next, but you want to attack those big pesky impediments first.

6. Negotiate all agreements, be it an agreement about velocity, schedule, cost even, what have you, but in the

back of your mind, you need to read page 77 in the Agile Practice Guide to understand the concept of an overarching master services agreement (MSA), An MSA with smaller little agreements is helpful. Read that in the Agile Practice Guide.

7. Have a mindset of collaboration in problem-solving,, seeking to understand first before being understood and that means using a method such as the D-I-G-C-I-V approach to break down the problem. Define the problem, identify the root cause, generate alternatives, choose the best alternative, implement that alternative, and verify that it actually worked.

8. As a servant leader, you must engage and support virtual teams. You should investigate the best way of engaging your team. Terms such as osmotic communication, information radiators, fishbowl window, should come to mind. Understanding concepts for not only collaborating but learning and mentoring are important. The concept of pair programming, understanding the team charter and pages 49 and 50 in your Agile Practice Guide will go a long way to help you.

9. Understand the importance of mentoring and promoting team performance by using emotional intelligence. These will help you immensely on your exam.

10. When it comes to the process piece from an Agile perspective, you need to also approach it from a soft-skills side even though it's Agile. Let me explain what I mean. When you think about executing a project from an Agile perspective, in the back of your mind should be people, value and business value. In order to execute this project with a bunch of processes, I should be doing it to deliver value to my client and I should be thinking; how I can deliver value in increments? Not in one big clump but in increments." When it comes to communications, your mindset and your thought process should be; how can I get the communication as expediently as possible to those who need it? My team members and stakeholders? Understand the concepts of osmotic communication, information radiators and co-located team.

11. When it comes to risk management, in the world of Agile, this is something that is built into the process through multiple iterations.

12. When it comes to engaging stakeholders, you must first think; the primary reason I'm doing this is success for my customer! So when you engage stakeholders, you're doing it with a purpose to succeed with your customer. Your strategy for engagement is all about what works best for the customer.

13. When it comes to planning and managing budget and resources, I would like you to pay close attention to page 92 in the Agile Practice Guide. It reads; "projects with high degrees of uncertainty may not benefit from detailed cost calculations due to frequent changes, instead, lightweight estimation methods can be used to generate a fast high-level forecast." So, your mindset should be, fast high-level forecasts are preferred from an Agile perspective because the scope hasn't been fully defined. Keep that in the back of your mind when answering Agile-related questions that may pertain to cost.

14. In the same token, as it pertains to schedule, you should remember that in the world of Agile, adaptive approaches are used and your short iterations or sprints are not going for years on end! At most, four weeks. So, these short iterations provide rapid

feedback on the approaches and suitability of deliverables and generally manifest as iterative scheduling and on-demand pull-based scheduling. Remember the importance of looking at pull-based systems instead of pushing work on people! That is also something to think about. Think about flow-based Agile, Kanban, iteration-based Agile, the world of Scrum, and understand how scheduling could be done in iterations and in adjusting the time-boxed approach or cadence, when it comes to flow-based Agile.

15. When it comes to quality, understand that quality is built into the processes of Agile. Within a Scrum framework, the constant iterations enable check-ins periodically, not just in sprint reviews. Also remember that sprint retrospectives are a perfect way of self-inspection within the team. That is also part of quality. So, it's important to remember quality in Agile is built-in. During retrospectives, the team regularly checks on the effectiveness of the quality process. They look at the root cause of issues then suggest trials of new approaches to improve quality.

16. When it comes to scope management from an Agile perspective, you should remember your product

backlog can change according to market conditions, business conditions and Enterprise Environmental Factors. Factors in the enterprise or factors that are in the marketplace could wildly affect your product backlog. Sometimes, external factors might implicitly eliminate so many of those stories in your backlog.

17. When it comes to integration from an Agile perspective, you should remember that the team is front and center, not the project manager because there is no project manager theoretically in the world of Scrum. So, in the world of Scrum, using it as an example, integration is done by the team.

18. Managing project changes is also important but remember if you want to change anything in the sprint backlog during the sprint, it has to go through the product owner and changes during the sprint are discouraged generally. Could they happen? They could but they must go through the product owner. Remember that changes to any part of the product backlog must be approved first of all by the product owner if it is going to be done.

With that said, anyone can add to the product backlog, but once the product backlog has been prioritized, and once you get an idea of what to be done in the next sprint, those ideas should be held in place as much as possible, and any changes must go through the product owner. It is a very important role. It is a role that should be respected. And if there's no respect for the product owner, there could be problems. You could get questions that really test your understanding of these roles.

I want to encourage you, if you have not already done so, do read page 40 - 41 in your Agile Practice Guide. Page 40 starts off with Agile roles.

19. In Agile, three common roles are used. It reads cross-functional team members, product owner, and team facilitator. Be sure to understand these three. The product owner works with stakeholders, customers, and the teams to define the product direction. Typically, product owners have a business background and bring subject matter expertise of a deep nature to the decisions. Sometimes the product owner requests help from people with deep domain expertise, such as architects or deep customer expertise, such as product

managers. In Agile, product owners create the backlog for and with the team, and the backlog helps the team see how to deliver the highest value without creating waste and for that reason, I often refer to the product owner as a Chief Value Officer.

20. A critical success factor for Agile teams is strong product ownership. Without attention to the highest value for the customer, the Agile team may create features that are not appreciated or otherwise insufficiently valuable, therefore wasting effort. So, in your big run-up to your exam, you should recognize these roles talked about on pages 40 and 41 of the Agile Practice Guide.

21. When it comes to planning and managing procurements, just remember those high-level Master Service Agreements and flexible arrangements! That's the way to go. Page 77 in the Agile Practice Guide is where you can read up about a lot of these contracts.

22. Remember in managing project artifacts, we have three artifacts in the world of Scrum: product backlog, sprint backlog, and Potentially Shippable Increment (PSI). But also remember that in a wider world of Agile, we might refer to other things as artifacts; storyboards, burn-up

charts, burn down, charts, Cumulative Flow Diagrams (CFDS), and so on. It is important again to assess how to execute a project; which approaches to use, which methodology, methods or practices, to go with. Which life-cycle or development approach to choose (iterative, incremental, Agile, or predictive) or even the possibility of a hybridized approach. What does that look like? Check out the Agile Practice Guide on pages 27 and 28. Understand those models for hybridization.

23. Understand that governance is not a bad word. It is the framework in which authority is exercised.

24. Understand the importance of managing issues from a Predictive standpoint, but when we talk about issues, a lot of times we're really using the word issue to replace impediment. As a good project manager, be proactive before a risk becomes an issue, you should have already thought about it. "What if this does indeed happen?"

25. Make sure that you're transferring knowledge so the project can continue. In the world of Agile, this is done daily and continuously; daily scrums, sprint reviews, backlog refinement, sprint retrospectives and sprint

planning. These are all vehicles for sharing knowledge and ensuring project continuity.

26. When you are closing out a project or phase, in the world of Predictive, it is different. In the world of Agile, we are ready to close at any time. In the sprint, remember what you are doing is iterative. You could be in a sprint and receive word it is the final sprint due to budgetary issues or other reasons. You should be able to close out the sprint expediently without having to go five sprints back to look for lessons learned because those lessons learned are not documented in the world of Scrum. In the world of Scrum, our retrospective is where we hold those conversations sacred. They are generally not shared outside of the team. They are not documented outside of the team either, so that knowledge is with the team and that's why as much as possible we do not like changing team members. We want to keep the team as intact as possible so that those good behaviors that have been learned, overcoming the five stages of team development and getting to the performing stage are kept intact across the team.

27. When we talk about Agile from a business perspective, you want to think about compliance. Compliance should also be thought about even in Agile, and these are things that could make their way into the product backlog.

28. When we talk about benefits and value, from the world of Agile in business, value trumps everything else. The product owner arranges those backlog items based on a few major parameters, largely value, urgency and risk. High-risk, high-value, do. Low-risk, high-value, do next. Low-value, low-risk, do last. High-risk, low-value, avoid because it does not make sense to do such high-risk/low-value items, although there may be exceptions due to stakeholder-preferences.

29. When it comes to evaluating the business environment from an Agile perspective, this is something the product owner and the team should be involved in doing but more so, the product owner. The product owner should continually review the business environment for impacts on the project and the product backlog. When it comes to understanding how organizational change impacts the project and vice-versa, all team members should have an awareness of

this, but most importantly the servant leader should be well aware of how to support organizational change.

And that is how you need to be thinking for your exam from an Agile perspective! Agile is huge on the exam but these guidelines will help you achieve success!

The CALM Mindset for a day to the exam

It has been said time and time again. What is my advice for you one day to the exam? My advice is very simple. You must be calm, cool, action-ready, a leader of yourself, and mindful. Let's expand on these just a little bit further.

C is for cool, confident, collected amidst other things. Let me explain these. When I say someone needs to be cool, I mean really relaxed. You need to be confident that you can do it, collected, composed, comfortable, and you need to cut off.

There's no point studying up till midnight and there is no point cramming. What you know, you know, but it's more in the mindsets that I've talked about all through this book that you need to focus on from an Agile perspective and even those permeate into predictive as well. So be cool confident, collected, and composed. Know that you've got this!

Cut-off 7:00 pm the night before. Watch a fun movie, eat a nice meal, call it a day until your exam.

A is for aware and action-ready. Things happen on the exam that will blow your mind. Students go into the exam thinking; "this is the day I will finish with the PMP"... only to discover the computer had a different idea! Some folks go into the test centers only to discover the machines are not acting right. Some folks start off taking the exam when all of a sudden, there's a power cut! What are you going to do? You need to be aware. You need to be action-ready. And if anything happens that requires you to take your exam on a different day, just remember, be cool. You're six feet above the ground, you have every reason to be happy. The exam is not the end of the world and that needs to be your mindset because when these things happen, people often think it's the end of the world but it's not. We need to remind ourselves to be aware, to be action-ready, and to be cool.

L – lead yourself, less is more. To be a good leader of yourself, you need to understand your brain is not a machine. You are not a machine. Less is more, stop cramming, stop forcing information into your head that cannot go in at the last minute, and have overarching mindset mantras instead. Instead of reading the entire Agile Practice Guide the night before, how about listening to the audio version of this book again and again. Or listening to my mindset YouTube videos/ Instead of you reading the PMBOK Guide over and over again, trying to cram 49 things that you should know or the domains and principles, focus on relaxing and understanding the big-picture and theme. Understanding *not* cramming is best. Less is more one day to the exam!

I would advise reading this over and over again and telling yourself you are a success, you are a champion, you are a victor in this battle towards this exam. That needs to be your mindset no matter what, no matter how you feel. When you're in the exam, I want you to remember I told you this; you need to be cool, calm, collected, aware, and a leader, and what does a leader do? A leader gives the people hope, so give yourself hope and lead yourself. Remember less is more.

M for be mindful! Be mindful of your resources.

Be mindful that second-guessing yourself is a poor choice. Don't do it. And that is the overarching framework for thinking when going into the exam. You've got to be calm. Remember my overarching mindsets.

The "PROJECT" Mindset

This is the PROJECT mnemonic for the success MINDSET.

P – *Problem Solver* is your name, problem-solving is your game! Always solve the real problem and move the situation forward through your actions.

R – *Respect Authority.* When you're answering questions on the exam, you need to be a problem solver, but you also need to respect authority. Do not go beyond your stated authority, respect authority, and don't disrespect authority.

O - is for *Own the Problem*, do not pass the blame.

J - is for *Just do what is required*, no gold plating.

E – *Equip the team*, mentor, train, coach. There's an expectation that you do this. Escalate as appropriate.

C – *Changes are important but review and check impact* before doing. If a change has been approved then it will be done, but when a change request comes, you should always review it.

Do an impact analysis with the team, get approval from the Change Control Board (including the sponsor and the customer if they are not on the board).

T – the final one is Take responsibility. Be accountable and show servant leader qualities.

And that is the project acronym for your exam. It's all in the thinking. You must own the problem; you must move that situation forward. Saying "no" to someone is not the best answer because it doesn't move the situation forward, so instead take responsibility and be a great leader.

Summarizing Agile thinking one more time, value people, transparent communications, positivity, likeability and not using punishment power. Understand the cross-functional team and the T-shaped/paint-drip/broken comb profile preference in agile. Value working with the customer daily. Understand that customer-requested change is not bad, it's good. Lastly, empower the team!

CHAPTER FIVE: PREDICTIVE PRACTICES

L et's talk about the world of Predictive at a very high level. You will get questions of a Predictive nature that may fall into the initiating process group.

For this, you want to know your project charter components and your overall project risks. You also want to understand that the high-level requirements and overall project risks are indeed part of your project charter. You also want to remember that the preliminary scope comes before the project. In other words, your customer before signing the dotted line sends out an RFP that has some preliminary scope details that must come before the project. It is a strange term but be aware that it does exist.

Now when you hear the term new project, sometimes that term new project could mean this is something we're thinking of doing. It does not immediately mean this is an authorization. When you hear the term new project, read between the lines and deduce if the project manager has already been assigned. If yes, that could mean there's already a charter in place, but if it says you are being considered as someone who could manage this new project and that project has not been authorized, read carefully. Be very comfortable when it comes to initiating, understanding that the project charter authorizes the project. Before the project charter is authorized, there are some pre-initiating events on page 30 in the PMBOK® Guide Sixth Edition. You got to understand the business case.

The business case and the benefits management plan are important. Understand that an output of developed project charter is the assumptions log. The assumption log we use that for all manner of assumptions. Schedule assumptions such as lead time could be in there. Understand that facilitation is important, it's the name of the game in initiating. You even come across some conflict even way early and initiating.

Understand that meeting management is important. Stakeholder identification could come in cycles before the project charter is authorized. And that's why one of the outputs within the project charter we call it a key stakeholder list. This is not a project charter, this is not a stakeholder register, it's a key stakeholder list. The stakeholder register should be updated so going into identifying stakeholders and initiating just remember you have your project charter that goes in to identify stakeholders with the key stakeholder list. We build on that stakeholder list. That's how the stakeholder register is created.

1. Going into planning, know your risk strategies. A team avoid, transfer, escalate, avoid, mitigate avoid, transfer, escalate, accept, mitigate. The two ways avoid, accept. Understand your positive strategies so things such as the E-A-S-E-E – Escalate, Accept, Share, Exploit, Enhance, EASEE. Know those two, know your early start and late start relationships and all of those dependencies. Understand that firm-fixed-price is favored a lot over time and materials when it comes to Predictive projects. Your customer wants to know what they are likely to pay in total.

2. When it comes to the aspect of executing, understand that interpersonal and soft skill use is better than relying on tools. Also, it is emphasized, questions will test your understanding of interpersonal and soft skills even from a Predictive standpoint that is why a lot of people are tricked into thinking they are in Agile questions because they are very interpersonal and soft-skills heavy. Those are not Agile questions, instead, you may realize some of those are Hybrid questions.

3. Monitoring and controlling, always follow recommended change control procedures. Understand you have no control over people's ability to request changes. Changes are not bad. Understand that variance analysis is used in monitoring and controlling.

4. And when it comes to closing, page 123 in the PMBOK® Guide is important. Follow the recommended closure procedures, administrative closure, contract closure, closing out documents releasing resources, releasing the team, understanding why a project was terminated and things such as that are all important.

CHAPTER SIX: KNOWLEDGE AREAS IN AGILE LENS

Likely, you are aware that the PMBOK® Guide Seventh edition has been announced as a reference for the PMP® exam. Yes!. But we need to go back to the basics and the basics cannot be found to the degree you need in the Seventh Edition.

For that reason, you find in the 7th mention is still made of the process groups and the language in the domains is pretty much knowledge area language.

Now the exam, you could look at the exam in two different ways. One is an Agile and Hybrid perspective, and the other is a largely Predictive perspective. However, this book, it cuts across both of those mindsets. What do I mean? Follow me to the fourth chapter of the PMBOK® Guide Sixth Edition. In the very beginning, in the opener to Chapter 4, on page 72, it states; "Key concepts for project integration management." Project integration management is specific to project managers. Whereas, other knowledge areas may be managed by specialists, see that? This area is one the PM cannot delegate. When they are in a Predictive environment or an environment where their role is preeminent, there is a mindset. That's true if you are tackling these Predictive questions. The idea is that the project manager cannot delegate this in a Predictive setting, it's their job.

Now talking about an Agile setting, if you follow me to page 74, it states at the bottom of the page, "Considerations for Agile and Adaptive environments." Iterative and Agile approaches promote the engagement of team members as local domain experts in integration management. The team members determine how plans and components should integrate. So, and in a Predictive world, there's a way things work. In an Agile world, as you can see page 74, it maps back to stuff that is in this book, believe it or not. What is on page 74?

Page 74 in the PMBOK® Guide Sixth Edition maps back to

page 91. So, if you look at page 91 in your Agile Practice Guide where it talks about integration management and its application in the Agile work process, it is a direct lift in the sixth edition if you didn't already know.

So, what am I trying to tell you, you need to read concepts trends tailoring and consideration for Agile for every process. That is the most important thing to do first because it helps you understand how this is done in the world of Predictive. How is it done in the world of Agile? See integration is done differently? Scope is done differently. The schedule is done differently, and that is part of the mindset I want you to get. The mindset of how knowledge areas should be practiced. I'm going to give you a very quick tour. Are you ready? Okay for the emphasis we will give a befitting background, where we touch on each knowledge area.

1. Integration: In the integration management knowledge area, it is crucial to understand and embrace your responsibility as a project manager in the Predictive approach. However, in Agile, integration becomes a team responsibility. It is important to note that the expectations of the project manager, as outlined in the PMBOK® Guide,

remain unchanged in an Adaptive environment. However, the detailed product planning and delivery control are delegated to the team. This collaborative approach emphasizes the project manager's accountability while involving the team in a more participative manner.

2. Scope: Moving on to the scope area, the Agile Practice Guide (page 91) highlights that Agile methods intentionally spend less time on defining and agreeing upon scope during the early stages of the project. Instead, they focus more on establishing a process for ongoing discovery and refinement. In an Agile environment, the scope evolves over time, and the product owner plays a significant role in prioritizing, organizing, and refining the scope based on value. This approach emphasizes the dynamic and iterative nature of Agile projects.

3. Schedule: Let's now discuss schedule management. In the Predictive world, a comprehensive schedule is typically developed, including tasks, subtasks, and milestones. However, according to page 92 of the Agile Practice Guide, Adaptive approaches employ shorter cycles to carry out work. Agile projects may utilize roadmaps at a high-level to define future plans, especially for large initiatives in large organizations. At lower levels, Agile projects rely on

sprint planning and a sprint backlog. It is important to note that the role of the project manager remains unchanged regardless of the development life cycle being used. However, to succeed in an Adaptive environment, the project manager must be familiar with the relevant tools and techniques.

4. Cost: Let's move on to cost budgeting in an Agile context. In Agile projects, the team size is typically kept constant, although team members may sometimes come and go. According to the Agile Practice Guide (page 92), projects with high degrees of uncertainty or undefined scope may not benefit from detailed cost calculations due to frequent changes. In an Adaptive environment, it is unnecessary and impractical to break down the cost calculations into detailed levels, as done in Predictive projects. Instead, scope and schedule adjustments are more commonly made to stay within the cost constraints. Agile projects often utilize lightweight estimation methods to quickly forecast labor costs. Flexibility and adaptability are crucial in managing cost in an Agile project.

5. Quality: Moving on to quality management, the perception of quality remains consistent in an Agile environment. Quality is defined as fitness for use, conformance to

requirements, and customer satisfaction. Agile methods emphasize frequent quality and review steps throughout the project, rather than leaving them towards the end. Regular retrospectives are conducted to assess the effectiveness of the quality process. Agile projects focus on small batches of work, incorporating as many project deliverables as possible, to identify and address inconsistencies and quality issues early on in the project life cycle when changes are less costly. While Agile projects still value traditional quality tools such as affinity diagrams, flowcharts, Ishikawa diagrams, histograms, scatter diagrams, and control charts, the emphasis on these tools may have reduced in recent exams.

6. Resource Management: Let's discuss resource management in an Agile environment. Projects with high variability benefit from team structures that promote focus and collaboration, such as self-organizing teams and individuals with T-shaped skills. Collaboration is highly valued in Agile projects, as it contributes to their success. Although collaboration is beneficial in other project environments, it is particularly critical in projects with a high degree of variability. In contrast, the world of Predictive projects may not emphasize co-located, cross-

functional, self-organized teams as strongly. In Predictive projects, the project manager often takes on the role of inspiring and motivating the team, whereas in Agile, collaboration and self-organization are key principles that drive the team's success.

7. Communications Management: In an Agile environment, effective communication is paramount. The Agile Practice Guide emphasizes the need for communication to flow continuously and copiously. The project environment in Agile projects is often subject to ambiguity and change, necessitating frequent communication to relay evolving and emerging details. To facilitate this, Agile teams streamline team members' access to information, conduct frequent team checkpoints, and promote co-location. Additionally, Agile methods encourage posting project artifacts transparently and holding regular stakeholder interviews to promote communication with management and stakeholders. The goal is to ensure timely and productive discussions and decision-making throughout the project.

8. Risk Management: Risks are uncertainties that can impact a project positively or negatively. It is important to note that risk management is not exclusive to the Predictive

approach; Agile projects also have their ways of managing risks. In high variability environments, which inherently entail more uncertainty and risk, Adaptive approaches leverage frequent reviews of incremental work products and cross-functional project teams to accelerate knowledge sharing and ensure effective risk understanding and management. Risks are considered when selecting iteration content, and the product owner should actively think about the risks associated with each story, not just its value. Agile projects require making informed decisions based on risk and value considerations, even if it means not pursuing high-risk, low-value items. The ability to make such decisions and address risks proactively is essential.

9. Procurement Management: In Agile environments, specific sellers may be engaged to extend the team's capabilities. This collaborative working relationship can lead to a shared risk procurement model, where both the buyer and the seller share the project's risks. It is recommended to familiarize oneself with Agile procurements and contracts, as understanding flexible contract options aligns with the principles of customer collaboration emphasized in the Agile manifesto.

10. Stakeholder Management: In projects experiencing a high degree of change, active engagement and participation with stakeholders are crucial to facilitate timely and productive discussions and decision-making. Adaptive teams directly engage with stakeholders, bypassing layers of management, to accelerate the sharing of information within and across the organization. Agile methods promote aggressive transparency by inviting stakeholders to project meetings and reviews or posting project artifacts in public spaces. This transparent approach helps quickly identify and address misalignment, dependencies, and other issues related to the changing project. Collaborative stakeholder engagement and open communication are key principles in Agile stakeholder management.

When preparing for the PMP® exam, it is crucial to adopt a collaborative mindset and emphasize teamwork rather than working in isolation. Problem-solving abilities are also key. To approach the exam effectively, ask yourself which phase you are in—initiating, planning, executing, monitoring and controlling, or closing—and what your current task entails.

Many questions on the exam will focus on managing change,

risks, issues, or problems. It is important to differentiate between these concepts. Risks have the potential to alter your approach, while issues may require adjustments to your plan or the need for workarounds. Additionally, the exam may test your ability to identify specific areas that require change, such as the schedule, cost, or scope, in response to new regulations or requests.

In cases involving regulations, adherence becomes mandatory, and changes should not be made without proper authorization. Practical questions are common on the exam, presenting scenarios where you, as a project manager, need to determine the appropriate course of action.

Always prioritize problem-solving and collaboration, and don't shy away from addressing unrealistic expectations from stakeholders. If stakeholders propose unrealistic dates, it is essential to review the situation with them and propose a more feasible approach. Open and transparent communication is key, and if someone is being dishonest, it is necessary to address the issue.

Considering the knowledge areas, integration management

revolves around making things happen and serving as the catalyst for integration. In scope management, unauthorized changes should be stopped, and authorized changes should go through proper change control procedures. In schedule management, maintaining control and building schedules correctly, considering artifacts like product backlogs and sprint backlogs, are important. Budget management emphasizes being a caring and respectful steward, aligning with the principles of the PMBOK® Guide Seventh Edition. Quality management emphasizes technical excellence and avoiding shortcuts.

Regarding resource management, remember that team members are humans first, and adopting a servant leadership approach is essential. Be an impartial bridge builder, connecting teams and helping them understand each other's work to overcome silos. In communications management, effective communication is vital, even if it requires manual processes in predictive environments.

The sender-receiver model remains relevant, whether communicating within the team or with wider stakeholders. Risk management in Agile projects is characterized by an inherent risk coping mechanism through iterative processes like inspecting and adapting. Intentionally addressing risks from various perspectives, such as mitigation, transfer, or enhancement, is crucial. Understanding the five strategies for negative risks/threats and positive risks/opportunities is important.

In procurement management, collaboration takes precedence over contract negotiation. When it comes to stakeholder management, focus on bridging gaps between stakeholders, engaging them actively, ensuring their satisfaction, and effectively solving their problems.

By adopting these mindsets and strategies, you will be better prepared for the PMP® exam and capable of approaching different scenarios with confidence and competence.

CHAPTER SEVEN: AGILE MINDSET FOR PMP® STUDENTS

Finally, let's conclude with some additional AGILE mindset principles I created specifically for PMP® Exam success!

1. **Advocate for agile principles, be a model for those principles, and lead the way**: This suggests that to be successful in an agile environment, one must not only understand and espouse the agile principles but also embody them in action. A leader in an agile setting should be an exemplar, showing through their actions how agility can improve project outcomes. They should be ready to experiment, adapt, and learn, demonstrating that these behaviors bring higher results than sticking strictly to a predefined plan.

2. **Ensure everyone in the firm has a common understanding of the agile manifesto values and principles**: Agile methods require a collective understanding and approach. They aren't just processes to be followed; they're a mindset that must be adopted by everyone involved. Therefore, ensuring that all members of the organization understand the values and principles of the Agile Manifesto is crucial for a successful transition to agile ways of working. This can result in more effective communication, collaboration, and problem-solving.

3. **Be a change agent at all levels and educate, mentor, and coach as soon as you find an opportunity to do so**: Agile champions or leaders should take every opportunity to educate and mentor others about agile practices, providing guidance and support. Change can be hard for many people, and it's important to remember that this transition can be a journey. Encouraging continuous learning and growth within the organization will help embed the agile principles more deeply and widely.

4. **Practice visualization using an information radiator**: This principle is about increasing transparency and fostering collaboration. Information radiators, such as Kanban boards or Burndown charts, visually represent the progress of work, thus

encouraging open communication. By making this information readily available, everyone can see the status of various tasks, understand the team's current workload, and identify potential bottlenecks or issues.

5. **Build trust and make the environment safe for the team**: A cornerstone of the agile philosophy is a high-trust environment where team members feel safe to express ideas, take risks, and admit mistakes. Fostering this kind of culture encourages transparency, learning, and innovation. Leaders must make it clear that failures are seen as opportunities to learn and grow, rather than something to be punished.

6. **Find new ways of working and experiment with new approaches to product development and project management**: This principle emphasizes the importance of a growth mindset in agile methods. Rather than adhering rigidly to one approach, teams should be encouraged to experiment, iterate, and learn from these experiences. This might mean trying different ways of working, new techniques, or tools, and continuously refining these based on what works best.

7. **Encourage knowledge sharing and collaboration to solve problems**: Agile teams are built on

collaboration and shared knowledge. By fostering a culture where everyone is encouraged to share their skills, insights, and ideas, teams can come up with more innovative solutions to problems. Remember, "none of us is as smart as all of us."

8. **Empower the team and allow self-organization, emergent leadership, self-leadership, and management**: In agile frameworks, teams are given the autonomy to manage their own work and make their own decisions. This fosters a sense of ownership, accountability, and motivation. Leaders should trust their teams and avoid micromanagement, allowing for emergent leadership where anyone can step up and lead depending on the task or situation at hand.

9. **Be a servant leader**: This means that a leader's primary role is to serve the team - helping them to perform at their best, removing obstacles in their path, and providing support and guidance as needed. This might mean adopting various roles depending on the team's needs - being a mentor, facilitator, conflict resolver, or motivator. The focus is on fostering a healthy and productive team environment rather than asserting authority.

10. **Deliver value to the customer**: This is a fundamental principle of the agile philosophy. The primary measure of success is delivering valuable software to the

customer. If a feature or task doesn't add value to the customer, it's considered waste and should be eliminated. This requires a deep understanding of customer needs, and a commitment to frequent and early delivery of valuable software.

11. **Engage the stakeholder**: Understanding that success is subjective and lies in the eye of the stakeholder is essential. Frequent and transparent communication with stakeholders is key to ensuring that their needs and expectations are being met. This means involving them in the planning and review processes, and keeping them informed about progress and changes.

12. **Plan adaptively**: Agile methods are iterative and adaptive, as opposed to traditional predictive planning. This means that instead of trying to plan everything out in detail from the start, teams should plan for the near term in detail and adapt as they go along, based on what they learn from each iteration. This allows for more flexibility to respond to change, and for continuous improvement towards excellence.

Remember, these principles are not just rules to be followed but reflect a mindset change that encourages flexibility, collaboration, and customer focus. They are guideposts to help navigate the often complex and uncertain landscape of product development and project management.

CHAPTER EIGHT: SHORT DOMAIN TESTS

PEOPLE DOMAIN QUESTIONS

Q1: You are the Scrum Master on a team of 7. When two parties are in conflict, the most productive approach to resolving the issue is to:
A) Do everything to compromise or reconcile
B) Focus on the differences
C) Smooth or accommodate while relying on a third-party mediator
D) Seek to understand the other's perspective

Q2: When leading a team, the most effective approach to ensure success is to:
A) Rely on a single leadership style to keep it fair
B) Strictly enforce rules and regulations to keep it uniform
C) Ignore individual needs and differences while sticking to your principles
D) Consider the needs and differences of all team members

Q3: You are a servant leader on a medium-sized project. When supporting team performance, the most effective approach is to:
A) Focus solely on individual performance
B) Use a one-size-fits-all approach
C) Ignore team member feedback
D) Appraise team performance and provide feedback

Q4: You are a project sponsor for a small engineering project. When empowering team members and stakeholders, the most effective approach is to:
A) Assign tasks without considering individual strengths
B) Refuse to delegate any authority or decision-making power
C) Use organizational management processes and restrict power
D) Organize around team strengths and bestow decision-making authority

Q5: You are the project manager on a hybrid endeavor in your company. When ensuring team members and stakeholders are adequately trained, the most effective approach is to:
A) Rely solely on online training in this current environment
B) Ignore individual training needs and focus on company objectives
C) Provide generic training materials
D) Determine required competencies and allocate resources for training

Q6: You have been asked to assemble a team for a medium-sized I.T endeavor. When building a team, the most effective approach is to:
A) Refuse to consider individual skills and abilities
B) Assign tasks without considering team dynamics
C) Ignore the importance of a positive team culture
D) Appraise stakeholder skills

Q7: What is the best approach to addressing impediments that arise during an agile project?
A) Ignore the impediment and see if it is removed by stakeholders on the team
B) Document the impediment using a risk register or project issue log
C) Identify the root cause and develop a plan to mitigate or eliminate it
D) Address the impediment with a workaround through the risk analysis process4

Q8: How can a Scrum team best ensure that story points are negotiated effectively?
A) Utilize a consensus-based approach
B) Outline the criteria for successful negotiation
C) Set a timeline for negotiation
D) Establish a win-lose negotiation strategy

Q9: What strategies can be used to build trust and influence stakeholders to accomplish project objectives?
A. Develop clear communication protocols
B. Draw up a negotiation agreement
C. Establish a collaborative environment
D. Foster positive relationships for self-interests

Q10: What is the most effective way to ensure shared understanding and build consensus when addressing potential misunderstandings in an Agile environment?
A. Survey only key parties
B. Investigate misunderstandings after the fact
C. Break down situation to identify the root cause of a misunderstanding
D. Support outcome of parties' agreement, good or negative

Q11: Which of the following actions can be taken to support virtual teams?
A. Examine virtual team member needs and investigate alternatives for engagement
B. Implement virtual team member engagement options
C. Continuously evaluate virtual team member engagement
D. All of the above

Q12: Which of the following is the key element of defining team ground rules in a project management context?
A. Communicating organizational principles with team and external stakeholders
B. Implementing advanced tools and technologies for project management
C. Focusing only on the technical aspects of project management
D. Delegating tasks to team members without proper guidance

Q13: What is the role of the Scrum Master in mentoring relevant stakeholders in a Scrum environment?
A. To allocate the time to remote pairing and recognize pairing opportunities
B. To facilitate and coach the team through Agile and Scrum concepts
C. To manage and rectify any violations of the team's ground rules
D. To communicate the team's progress and success to external stakeholders

Q14: What is the role of emotional intelligence in promoting team performance in a project environment?

A. Identifying and addressing conflicts within the team using your expert judgment to come up with solutions for involved parties

B. Assessing behavior through the use of personality indicators and adjusting to the emotional needs of key project stakeholders

C. Implementing strict rules and consequences for non-compliant team members who may violate social contracts

D. Prioritizing tasks based solely on objective measures and disregarding team morale and motivation

PEOPLE DOMAIN ANSWERS

Q1: You are the Scrum Master on a team of 7. When two parties are in conflict, the most productive approach to resolving the issue is to:
A) Do everything to compromise or reconcile
B) Focus on the differences
C) Smooth or accommodate while relying on a third-party mediator
D) Seek to understand the other's perspective

Answer: D) Seek to understand the other's perspective.
Rationale: Being open to understanding the other party's perspective can be the most productive approach to resolving a conflict, as it allows both sides to gain a better understanding of the situation and can help to identify common ground and potential solutions.

Q2: When leading a team, the most effective approach to ensure success is to:
A) Rely on a single leadership style to keep it fair
B) Strictly enforce rules and regulations to keep it uniform
C) Ignore individual needs and differences while sticking to your principles
D) Consider the needs and differences of all team members

Answer: D) Consider the needs and differences of all team members.
Rationale: Taking the time to consider the needs and differences of all team members is the most effective approach to ensure success when leading a team, as it allows the leader to gain a better understanding of each individual's perspective and can help to create a better team dynamic and foster collaboration.

Q3: You are a servant leader on a medium-sized project. When supporting team performance, the most effective approach is to:
A) Focus solely on individual performance
B) Use a one-size-fits-all approach
C) Ignore team member feedback
D) Appraise team performance and provide feedback

Answer: D) Appraise team performance and provide feedback.
Rationale: Appraising the team's performance and providing tailored feedback to individual team members is the most effective approach when supporting team performance, as it allows the leader to gain a better understanding of

each individual's performance and can help to identify areas for improvement and growth.

Q4: You are a project sponsor for a small engineering project. When empowering team members and stakeholders, the most effective approach is to:
A) Assign tasks without considering individual strengths
B) Refuse to delegate any authority or decision-making power
C) Use organizational management processes and restrict power
D) Organize around team strengths and bestow decision-making authority

Answer: D) Organize around team strengths and bestow decision-making authority. Rationale: Rationale: Organizing a team around the strengths of its members and bestowing decision-making authority is the most effective approach when empowering team members and stakeholders, as it allows the leader to gain a better understanding of each individual's capabilities and can help to create an environment of mutual respect and collaboration.

Q5: You are the project manager on a hybrid endeavor in your company. When ensuring team members and stakeholders are adequately trained, the most effective approach is to:
A) Rely solely on online training in this current environment
B) Ignore individual training needs and focus on company objectives
C) Provide generic training materials
D) Determine required competencies and allocate resources for training

Answer: D) Determine required competencies and allocate resources for training.
Rationale: Determining the required competencies and allocating resources for training is the most effective approach when ensuring team members and stakeholders are adequately trained, as it allows the leader to gain a better understanding of each individual's training needs and can help to ensure that the training is tailored to their needs.

Q6: You have been asked to assemble a team for a medium-sized I.T endeavor. When building a team, the most effective approach is to:
A) Refuse to consider individual skills and abilities
B) Assign tasks without considering team dynamics
C) Ignore the importance of a positive team culture
D) Appraise stakeholder skills

Answer: D) Appraise stakeholder skills.

Rationale: Beware of instances where the BEST answer is not expanded upon. Appraising stakeholder skills, analyzing team dynamics, and planning team tasks is the most effective approach when building a team, as it allows the leader to gain a better understanding of each individual's skills and abilities and can help to create a positive team culture and ensure that tasks are assigned appropriately.

Q7: What is the best approach to addressing impediments that arise during an agile project?
A) Ignore the impediment and see if it is removed by stakeholders on the team
B) Document the impediment using a risk register or project issue log
C) Identify the root cause and develop a plan to mitigate or eliminate it
D) Address the impediment with a workaround through the risk analysis process

Answer: D) The best approach to addressing impediments during an agile project is to identify the root cause and develop a plan to mitigate or eliminate it.
Rationale: Impediments are not kept in a risk register, mitigation and elimination are risk responses (generally speaking). Risk analysis is not the approach for impediments.

Q8: How can a Scrum team best ensure that story points are negotiated effectively?
A) Utilize a consensus-based approach
B) Outline the criteria for successful negotiation
C) Set a timeline for negotiation
D) Establish a win-lose negotiation strategy

Answer: A)
Rationale: A consensus-based approach is the best way to ensure that story points are negotiated effectively because it allows all members of the team to have input on the negotiation process and reach an agreement that everyone is comfortable with.

Q9: What strategies can be used to build trust and influence stakeholders to accomplish project objectives?
A. Develop clear communication protocols
B. Draw up a negotiation agreement
C. Establish a collaborative environment
D. Foster positive relationships for self-interests

Answer: C. Establish a collaborative environment
Rationale: Establishing a collaborative environment is key to building trust and influencing stakeholders to accomplish project objectives. Clear communication protocols and positive relationships can support this goal but are not adequate on their own. Negotiation agreements are not necessary in this scenario. The Agile manifesto espouses collaboration over negotiation!

Q10: What is the most effective way to ensure shared understanding and build consensus when addressing potential misunderstandings in an Agile environment?
A. Survey only key parties
B. Investigate misunderstandings after the fact
C. Break down situation to identify the root cause of a misunderstanding
D. Support outcome of parties' agreement, good or negative

Answer: C. Break down situation to identify the root cause of a misunderstanding
Rationale: It is essential to break down a situation to properly identify the root cause of a misunderstanding before attempting to build consensus. Surveying all necessary parties and investigating potential misunderstandings can help to do this, but breaking down the situation is the most effective way to ensure shared understanding and build consensus.

Q11: Which of the following actions can be taken to support virtual teams?
A. Examine virtual team member needs and investigate alternatives for engagement
B. Implement virtual team member engagement options
C. Continuously evaluate virtual team member engagement
D. All of the above

Answer: D. All of the above
Rationale: To engage and support virtual teams, it is important to first examine the needs of virtual team members, such as environment, geography, culture, and global considerations. Next, alternatives for virtual team member engagement, such as communication tools and colocation, should be investigated. Finally, options for virtual team member engagement should be implemented and the effectiveness of the engagement should be continually evaluated.

Q12: Which of the following is the key element of defining team ground rules in a project management context?

PHILL AKINWALE, PMP, OPM3

A. Communicating organizational principles with team and external stakeholders
B. Implementing advanced tools and technologies for project management
C. Focusing only on the technical aspects of project management
D. Delegating tasks to team members without proper guidance

Answer: A. Communicating organizational principles with team and external stakeholders
Rationale: Defining team ground rules is an important aspect of project management, particularly when it comes to Agile methodologies. The key element of defining team ground rules is to communicate the organizational principles with the team and external stakeholders. This helps to establish a common understanding of the ground rules and ensure that everyone is on the same page in terms of the expectations and behaviors required for the project's success. By doing so, the project manager can create an environment that fosters adherence to the ground rules, leading to better collaboration and teamwork.

Q13: What is the role of the Scrum Master in mentoring relevant stakeholders in a Scrum environment?
A. To allocate the time to remote pairing and recognize pairing opportunities
B. To facilitate and coach the team through Agile and Scrum concepts
C. To manage and rectify any violations of the team's ground rules
D. To communicate the team's progress and success to external stakeholders

Answer: B. To facilitate and coach the team through Agile and Scrum concepts
Rationale: The role of the Scrum Master in a Scrum environment is to facilitate and coach the team, and to ensure that the team is working effectively. As a mentor, the Scrum Master should allocate time to mentoring and be proactive in recognizing mentoring opportunities, in order to support and develop the team members and help them reach their full potential. Option B is a general description of the Scrum Master's role, while options A, C and D are not specific to mentoring.

Q14: What is the role of emotional intelligence in promoting team performance in a project environment?
A. Identifying and addressing conflicts within the team using your expert judgment to come up with solutions for involved parties
B. Assessing behavior through the use of personality indicators and adjusting to the emotional needs of key project stakeholders
C. Implementing strict rules and consequences for non-compliant team members who may violate social contracts

D. Prioritizing tasks based solely on objective measures and disregarding team morale and motivation

Answer: B. Assessing behavior through the use of personality indicators and adjusting to the emotional needs of key project stakeholders
Rationale: Promoting team performance through the application of emotional intelligence involves assessing behavior through the use of personality indicators and adjusting to the emotional needs of key project stakeholders. This helps to create a positive project environment that considers the emotional needs of team members, leading to better teamwork and overall performance.

PROCESS DOMAIN QUESTIONS

1. PLI, a web design company, is currently working on a project to redesign a client's e-commerce website. The project has a tight deadline of four months, and the team is working hard to deliver the project on time. Leroy is the project manager, and Angela is the project sponsor. When executing a web design project, which approach should the project team use to deliver business value incrementally?
a. Implement all project requirements simultaneously
b. Develop the most complex project requirements first
c. Subdivide project tasks to find the minimum viable product
d. Implement the easiest project requirements first

2. PLI, a web design company, is currently working on a project to develop a new website for a non-profit organization. The project has a budget of $100,000 and a timeline of six months. The project team is working in an Agile environment and following a Scrum framework. What should first be considered when determining communication methods for entities on this web design project?
a. The team's preferred communication methods
b. The level of detail required by stakeholders
c. The cost of the communication methods
d. The project timeline

3. PLI, a web design company, is working on a project to develop a new website for a client. The project has a several risks and a tight deadline of four months and a budget of $150,000. The project team is working in an Agile environment and following the Scrum framework. What is the purpose of iteratively assessing and prioritizing risks on this project?
a. To identify and address risks throughout the project
b. To develop a risk management plan for the project
c. To mitigate all project risks before project execution
d. To identify risks that have no impact on the project
4. During the project planning phase, the project team identifies the risk that the project may experience delays due to unforeseen technical issues. To manage this risk, the team decides to allocate additional resources and establish a contingency plan to address any technical issues that may arise during the project.

Which of the following best describes the risk management strategy
that the project team is implementing?
a. Risk avoidance
b. Risk transfer
c. Passive risk acceptance
d. Risk mitigation

5. PLI, a web design company, is working on a project to redesign a
client's e-commerce website. The project has a tight deadline of four
months and a budget of $100,000 and stakeholders with competing
self-interests. How should the project team engage stakeholders in a
web design project?
a. By analyzing stakeholders based on their power and influence
b. By communicating only with stakeholders who have a high level of
interest in the project
c. By ignoring stakeholders who are not directly impacted by the project
d. By communicating with all stakeholders, regardless of their level of
interest or impact on the project

6. During the project planning phase, the project team conducts a
stakeholder analysis to identify and categorize stakeholders based on
their power, interest, influence, and impact on the project. The team
identifies Jen and Shouna as high-power, high-interest stakeholders
who have a significant impact on the project. Which of the following
best describes the strategy that the project team should use to engage
Jen and Shouna?
a. Keep Jen and Shouna informed of project updates and progress
throughout the project
b. Minimize communication with Jen and Shouna to avoid delays in the
project
c. Only engage Jen and Shouna during the project planning and closing
phases
d. Ignore Jen and Shouna's concerns and focus on delivering the project
on time and within budget

7. PLI, a web design company, is working on a project to develop a new
website for a non-profit organization. The project has a tight budget of

$50,000 and a deadline of six months. How can a project team anticipate future budget challenges in a web design project?
a. By increasing the project budget from the start to prevent future increases
b. By tracking and monitoring budget variations throughout the project
c. By reducing the scope of the project to prevent changes or overruns
d. By avoiding project risks by taking strict change control and mitigation actions

8. Leroy is the project manager, and Angela is the project sponsor of a web-design project. Halfway through the project, the project team realizes that they have overspent their budget by $5,000 due to unexpected scope changes and increased resource costs. They also anticipate that there may be additional budget challenges in the coming months. What is the best course of action for the project team to manage this budget overrun?
a. Ignore the budget overrun and focus on delivering the project on time
b. Reduce project scope to bring the project back within budget since the team is able to discuss and meet with customers
c. Use up the management reserves available to Leroy, the project manager since no permission is needed to do so
d. Implement cost-saving measures to bring the project back within budget

9. Praizion, a media company, is working on a project to create a media campaign for the Department of Commerce. The project has a tight deadline of three months and a budget of $500,000. The project team is working in a traditional environment and following the Waterfall methodology. Jane is the project manager, and John is the project sponsor. What is the purpose of utilizing benchmarks and historical data when estimating project tasks for this Hybrid project?
a. To ensure the project team meets their story point goals
b. To ensure the project is completed within the scheduled timeline
c. To reduce the project scope and budget
d. To manage the resources required for the project

10. Praizion, a media company, is working on a project to create a media campaign for the Department of Commerce. During the project planning phase, the project team determines the quality standard required for project deliverables and recommends options for improvement based on quality gaps. They also plan to continually survey project deliverable quality. What is the purpose of continually surveying project deliverable quality in a media campaign project?
a. To identify areas for improvement
b. To reduce project costs
c. To accelerate the project timeline
d. To increase the project scope

11. Halfway through the project, the project team realizes that there is a quality issue with the deliverables. The customer has expressed concerns about the quality of the work, and the project team has identified several areas for improvement. What is the best course of action for the project team to manage this quality issue?

a. Ignore the quality issue and continue with the project
b. Revise the project scope to exclude the areas with quality issues
c. Implement quality improvement measures to address the quality issue
d. Request additional funds from the project sponsor to cover the cost of rework

12. Praizion, a media company, is working on a project to create a media campaign for the Department of Commerce. The project has a tight deadline of three months and a budget of $500,000. During the project planning phase, the project team identifies that the project scope is unclear, and there is a risk of scope creep. What should the project team do?
a. Ignore the scope issue and continue with the project
b. Revise the project timeline to accommodate the scope issue
c. Clarify the project scope with the customer and stakeholders
d. Request additional funds from the project sponsor to cover the expanded scope

13. Praizion, a media company, is working on a project to create a media campaign for the Department of Commerce. What is the purpose of breaking down the scope of the media campaign project?
a. To reduce project costs
b. To increase the project scope
c. To identify project requirements and prioritize them
d. To accelerate the project timeline

14. Praizion, a media company, is working on a project to create a media campaign for the Department of Commerce. Why is it important to assess consolidated project plans for dependencies, gaps, and continued business value in the Praizion media campaign project?
a. To identify the resources required for the project
b. To identify gaps in the project budget and make plans as small as possible
c. To reduce the project scope and ensure it fits within the timeline
d. To ensure that the project aligns with business objectives

15. Praizion, a media company, is working on a project to create a media campaign for the Department of Commerce. Why is it important to determine a change response to move the project forward in a media campaign project?
a. To reduce the possibility of changes coming from the customer to shorten the project
b. To avoid all project risks which could lead to the project being completed prematurely
c. To ensure that the project is completed within the scheduled timeline
d. To handle changes effectively and efficiently

16. During the project planning phase, the project team identifies that they will need to procure additional resources, such as hardware and software, to complete the project. They estimate the resource requirements and needs, communicate the requirements to the project sponsor, and develop a plan to manage the procurement process. What is the purpose of managing suppliers/contracts in a media campaign project?

a. To reduce project costs in order to satisfy stakeholders financial concerns

b. To identify project risks to avoid any positive risks that may occur

c. To ensure that project deliverables meet the required quality standards

d. To accelerate the project timeline in order to satisfy stakeholders

17. PM Anonymous, a project management consulting firm, is managing a large-sized Agile IT project for a client. The project team has been efficiently trained on all tools and systems to work in an Agile environment. Halfway through the project, the Scrum Master realizes that some of the project artifacts are not up to date. Some team members have not been updating the Agile project management tool regularly, and there is confusion about the latest version of the code. What is the best course of action for the Scrum Master to manage this issue with project artifacts?

a. Ignore the issue and focus on completing the project on time

b. Schedule a team meeting to discuss the importance of updating the project artifacts

c. Train team members on how to use the Agile project management tool effectively

d. Implement a peer review process to ensure that all project artifacts are up to date

18. During the project initiation phase, the project manager meets with the client to discuss the project requirements and determine the appropriate project methodology. The client is interested in an Agile approach to project management, but the project team is more familiar with Waterfall methodology. The project manager considers the project's complexity, risk factors, and resource availability to determine the most appropriate project methodology for the project. After conducting a detailed analysis, the project manager recommends a hybrid project management methodology that combines elements of Agile and Waterfall methodologies. The project manager explains the hybrid methodology to the client and the project team and obtains their buy-in. What is the primary benefit of using a hybrid project management methodology for this project?

a. Enables the project team to use Agile practices to manage project uncertainty and complexity

b. Ensures that the project is completed within the established timeline and budget

c. Provides a clear and well-defined project plan with detailed deliverables and timelines

d. Enables the project team to use iterative and incremental development practices to improve project quality

19. What is the purpose of defining escalation paths and thresholds in a large-sized Agile IT project?

a. To ensure that project information is properly stored

b. To ensure that project information is secure

c. To define how and when project issues will be escalated

d. To ensure that project deliverables meet the required quality standards

20. PM Anonymous, a project management consulting firm, is managing a new project for a client. The project involves developing a new software application for the client's business. The project has a duration of twelve months, and the budget is $2,000,000. The project team includes ten developers, three business analysts, and one project manager. During the project planning phase, the project manager is tasked with defining the appropriate governance structure for the project and determining escalation paths and thresholds. What is the primary purpose of establishing a project governance structure?

a. To ensure that the project is completed within the established timeline and budget

b. To provide a clear and well-defined project plan with detailed deliverables and timelines

c. To ensure that the project is executed effectively and efficiently and that high-quality project deliverables are produced

d. To establish communication methods, channels, frequency, and level of detail for all stakeholders

21. What is the purpose of collaborating with relevant stakeholders on the approach to resolve project issues in a large-sized Agile IT project?
a. To ensure that project information is properly stored
b. To ensure that project information is secure
c. To identify areas for improvement and adjust the project as necessary
d. To address project issues in a collaborative and effective manner

22.   During the project execution phase, the project team encounters an issue with the software application's user interface. The client has raised concerns about the user interface, and the project team is struggling to address the issue. The project managers convene a meeting with the project team and stakeholders to discuss the issue and identify the root cause. After analyzing the issue, the project team realizes that the user interface issue is related to a coding error in the software application. The project team determines that they need to rewrite a significant portion of the code to address the issue. What is the best course of action for the project managers to manage this issue?
a. Ignore the issue and continue with the project as planned
b. Assign blame to the project team member responsible for the coding error
c. Schedule an emergency meeting with the client to discuss the issue and potential solutions
d. Collaborate with the project team to develop a plan to address the issue and communicate the plan to the client

23. What is the purpose of confirming the approach for knowledge transfers in a large-sized Agile IT project?
a. To ensure that project information is properly stored
b. To ensure that project information is secure
c. To ensure that project knowledge is transferred effectively and efficiently
d. To identify areas for improvement and adjust the project as necessary

24. What is the main benefit of conducting a knowledge transfer session at the end of a project?
a. Helps the project team members improve their skills and knowledge
b. Facilitates the transfer of knowledge from the departing team members to the remaining team members
c. Enables the project manager to evaluate the performance of the project team members
d. Provides an opportunity to celebrate the successful completion of the project

25.   What is the purpose of concluding activities to close out a large-sized Agile IT project, such as final lessons learned and retrospective?
a. To ensure that the project team has completed all the required tasks
b. To assess project performance and identify areas for improvement
c. To accelerate the project timeline in order to complete the scope
d. To reduce the project budget in order to return money to the cost-centers

26.   The project manager meets with the project team to discuss the closure plan and identify the specific tasks that need to be completed. The team determines that they need to conduct a final project review to evaluate the project's success, identify any areas for improvement, and document the lessons learned. They also need to archive the project documents and transfer the final deliverables to the client. What is the main benefit of conducting a final project review?
a. To evaluate the performance of the project team members and give appropriate rewards and recognition
b. To identify areas for improvement and apply the lessons learned to future projects
c. To celebrate the successful completion of the project
d. To determine if the project met the customer's requirements and have a final commendation from the stakeholders

PRAIZION PROCESS DOMAIN

Task 1: Execute project with the urgency required to deliver business value

1. PLI, a web design company, is currently working on a project to redesign a client's e-commerce website. The project has a tight deadline of four months, and the team is working hard to deliver the project on time. Leroy is the project manager, and Angela is the project sponsor. When executing a web design project, which approach should the project team use to deliver business value incrementally?

a. Implement all project requirements simultaneously
b. Develop the most complex project requirements first
c. Subdivide project tasks to find the minimum viable product
d. Implement the easiest project requirements first

Rationale: The correct answer is (c) - subdivide project tasks to find the minimum viable product. By subdividing project tasks, the team can identify the most important requirements that will deliver the most value to the business and customers. Customer feedback given about the MVP will also ensure that they are building the right thing.

Task 2: Manage communications

2. PLI, a web design company, is currently working on a project to develop a new website for a non-profit organization. The project has a budget of $100,000 and a timeline of six months. The project team is working in an Agile environment and following a Scrum framework. What should first be considered when determining communication methods for entities on this web design project?

a. The team's preferred communication methods
b. The level of detail required by stakeholders
c. The cost of the communication methods
d. The project timeline

Rationale: The correct answer is (b) - the level of detail required by stakeholders. When determining communication methods for "entities" on this web design project, it is important to consider the level of detail required by each stakeholder to ensure that project information is communicated effectively. The team's preferred communication methods are secondary compared to the stakeholder's because the communication

Task 3: Assess and manage risks

3. PLI, a web design company, is working on a project to develop a new website for a client. The project has a several risks and a tight deadline of four months and a budget of $150,000. The project team is working in an Agile environment and following the Scrum framework. What is the purpose of iteratively assessing and prioritizing risks on this project?

a. To identify and address risks throughout the project
b. To develop a risk management plan for the project

c. To mitigate all project risks before project execution
d. To identify risks that have no impact on the project

Rationale: The correct answer is (a) - to identify and address risks throughout the project. Iteratively assessing and prioritizing risks allows the project team to identify and address risks as they arise, ensuring that the project stays on track and meets its objectives. Not all risks are "mitigated". Some could be avoided, accepted, escalated or transferred for example.

Task 3: Assess and manage risks

4. During the project planning phase, the project team identifies the risk that the project may experience delays due to unforeseen technical issues. To manage this risk, the team decides to allocate additional resources and establish a contingency plan to address any technical issues that may arise during the project.

Which of the following best describes the risk management strategy that the project team is implementing?

a. Risk avoidance
b. Risk transfer
c. Passive risk acceptance
d. Risk mitigation

Rationale: This is a tricky question because technically, the allocation of additional resources is mitigation but the layer of contingency planning could be viewed as active risk acceptance. The correct answer is therefore (d) - risk mitigation. The project team is implementing a risk management strategy of risk mitigation, by allocating additional resources and establishing a contingency plan to address the risk of technical issues that may cause delays in the project. Risk mitigation aims to reduce the likelihood and/or impact of a risk on the project, and is a common risk management strategy in Agile projects

where the team is continuously assessing and managing project risks throughout the project lifecycle.

Task 4: Engage stakeholders

5. PLI, a web design company, is working on a project to redesign a client's e-commerce website. The project has a tight deadline of four months and a budget of $100,000 and stakeholders with competing self-interests. How should the project team engage stakeholders in a web design project?

a. By analyzing stakeholders based on their power and influence

b. By communicating only with stakeholders who have a high level of interest in the project

c. By ignoring stakeholders who are not directly impacted by the project

d. By communicating with all stakeholders, regardless of their level of interest or impact on the project

Rationale: The correct answer is (a) - by analyzing stakeholders based on their power and influence. Analyzing stakeholders based on their power and influence can help the project team to prioritize stakeholder engagement and ensure that the project meets stakeholder needs and expectations. Not every stakeholder needs the same level or type of communication. The best way to engage stakeholders is based on their power and influence.

Task 4: Engage stakeholders

6. During the project planning phase, the project team conducts a stakeholder analysis to identify and categorize stakeholders based on their power, interest, influence, and impact on the project. The team identifies Jen and Shouna as high-power, high-interest stakeholders who have a significant impact on the project. Which of the following best describes the strategy that the project team should use to engage Jen and Shouna?

a. Keep Jen and Shouna informed of project updates and progress throughout the project

b. Minimize communication with Jen and Shouna to avoid delays in the project

c. Only engage Jen and Shouna during the project planning and closing phases

d. Ignore Jen and Shouna's concerns and focus on delivering the project on time and within budget

Rationale: The correct answer is (a) - keep Jen and Shouna informed of project updates and progress throughout the project. Engaging high-power, high-interest stakeholders like Jen and Shouna is critical to the success of the

project. By keeping them informed of project updates and progress throughout the project, the project team can ensure that Jen and Shouna's concerns are addressed and that the project meets their expectations. This also enables the project team to maintain a good relationship with these stakeholders and increase their likelihood of success.

Task 5: Plan and manage budget and resources

7. PLI, a web design company, is working on a project to develop a new website for a non-profit organization. The project has a tight budget of $50,000 and a deadline of six months. How can a project team anticipate future budget challenges in a web design project?

a. By increasing the project budget from the start to prevent future increases
b. By tracking and monitoring budget variations throughout the project
c. By reducing the scope of the project to prevent changes or overruns
d. By avoiding project risks by taking strict change control and mitigation actions

Rationale: The correct answer is (b) - by tracking and monitoring budget variations throughout the project. By tracking and monitoring budget variations, the project team can anticipate future budget challenges and adjust the project budget and resources as necessary to ensure project success. It is not realistic to think about avoiding all project risks or preventing future increases/overruns.

Task 5: Plan and manage budget and resources

8. Leroy is the project manager, and Angela is the project sponsor of a web-design project. Halfway through the project, the project team realizes that they have overspent their budget by $5,000 due to unexpected scope changes and increased resource costs. They also anticipate that there may be additional budget challenges in the coming months. What is the best course of action for the project team to manage this budget overrun?

a. Ignore the budget overrun and focus on delivering the project on time
b. Reduce project scope to bring the project back within budget since the team is able to discuss and meet with customers
c. Use up the management reserves available to Leroy, the project manager since no permission is needed to do so
d. Implement cost-saving measures to bring the project back within budget

Rationale: The correct answer is (d) - implement cost-saving measures to bring the project back within budget. Overspending the budget is a common risk in projects, and it is essential to manage this risk effectively to ensure project

success. Implementing cost-saving measures, such as renegotiating contracts, reducing non-essential project activities, or re-allocating resources, can help the project team bring the project back within budget while minimizing the impact on project scope and timeline. Bear in mind that management reserves are not available to the project manager. Permission is needed to use them. There have been scope changes and this you cannot ignore.

Task 6: Plan and manage schedule

9. Praizion, a media company, is working on a project to create a media campaign for the Department of Commerce. The project has a tight deadline of three months and a budget of $500,000. The project team is working in a traditional environment and following the Waterfall methodology. Jane is the project manager, and John is the project sponsor. What is the purpose of utilizing benchmarks and historical data when estimating project tasks for this Hybrid project?

a. To ensure the project team meets their story point goals
b. To ensure the project is completed within the scheduled timeline
c. To reduce the project scope and budget
d. To manage the resources required for the project

Rationale: The correct answer is (b) - to ensure the project is completed within the scheduled timeline. Utilizing benchmarks and historical data can help the project team accurately estimate project tasks and ensure that the project is completed on time. There is no such thing as "story-point goals" in Agile. This could generate undesirable behavior.

Task 7: Plan and manage quality of products/deliverables

10. Praizion, a media company, is working on a project to create a media campaign for the Department of Commerce. During the project planning phase, the project team determines the quality standard required for project deliverables and recommends options for improvement based on quality gaps. They also plan to continually survey project deliverable quality. What is the purpose of continually surveying project deliverable quality in a media campaign project?

a. To identify areas for improvement
b. To reduce project costs
c. To accelerate the project timeline
d. To increase the project scope

Rationale: The correct answer is (a) - to identify areas for improvement. Continually surveying project deliverable quality allows the project team to identify areas where improvements can be made, ensuring that the project meets the required quality standards.

Task 7: Plan and manage quality of products/deliverables

11. Halfway through the project, the project team realizes that there is a quality issue with the deliverables. The customer has expressed concerns about the quality of the work, and the project team has identified several areas for improvement. What is the best course of action for the project team to manage this quality issue?

a. Ignore the quality issue and continue with the project

b. Revise the project scope to exclude the areas with quality issues

c. Implement quality improvement measures to address the quality issue

d. Request additional funds from the project sponsor to cover the cost of rework

Rationale: The correct answer is (c) - implement quality improvement measures to address the quality issue. Maintaining quality standards is a critical part of project management. Implementing quality improvement measures, such as revising the quality standard, conducting additional quality checks, or providing training to team members, can help the project team address the quality issue and prevent further quality issues. This enables the project team to deliver high-quality deliverables that meet or exceed the customer's expectations, thereby improving customer satisfaction and the likelihood of project success.

Task 8: Plan and manage scope

12. Praizion, a media company, is working on a project to create a media campaign for the Department of Commerce. The project has a tight deadline of three months and a budget of $500,000. During the project planning phase, the project team identifies that the project scope is unclear, and there is a risk of scope creep. What should the project team do?

a. Ignore the scope issue and continue with the project

b. Revise the project timeline to accommodate the scope issue

c. Clarify the project scope with the customer and stakeholders

d. Request additional funds from the project sponsor to cover the expanded scope

Rationale: The correct answer is (c) - clarify the project scope with the customer and stakeholders. Unclear project scope and the risk of scope creep can lead to project delays and overspending, ultimately impacting project success. Clarifying the project scope with the customer and stakeholders can help ensure that the project team has a clear understanding of the project requirements and can develop a plan to meet those requirements within the project timeline and budget. This enables the project team to deliver the project successfully and meet or exceed the customer's expectations.

Task 8: Plan and manage scope

    13. Praizion, a media company, is working on a project to create a media campaign for the Department of Commerce. What is the purpose of breaking down the scope of the media campaign project?

a. To reduce project costs
b. To increase the project scope
c. To identify project requirements and prioritize them
d. To accelerate the project timeline

Rationale: The correct answer is (c) - to identify project requirements and prioritize them. Breaking down the scope of a media campaign project allows the project team to identify the project requirements and prioritize them based on their importance to the project.

Task 9: Integrate project planning activities

    14. Praizion, a media company, is working on a project to create a media campaign for the Department of Commerce. Why is it important to assess consolidated project plans for dependencies, gaps, and continued business value in the Praizion media campaign project?

a. To identify the resources required for the project
b. To identify gaps in the project budget and make plans as small as possible
c. To reduce the project scope and ensure it fits within the timeline
d. To ensure that the project aligns with business objectives

Rationale: The correct answer is (d) - to ensure that the project aligns with business objectives. Assessing consolidated project plans for dependencies, gaps, and continued business value ensures that the project aligns with the overall business objectives and provides value to the organization.

Task 10: Manage project changes

15. Praizion, a media company, is working on a project to create a media campaign for the Department of Commerce. Why is it important to determine a change response to move the project forward in a media campaign project?

a. To reduce the possibility of changes coming from the customer to shorten the project

b. To avoid all project risks which could lead to the project being completed prematurely

c. To ensure that the project is completed within the scheduled timeline

d. To handle changes effectively and efficiently

Rationale: The correct answer is (d) - to handle changes effectively and efficiently. Determining a change response to move the project forward ensures that the project team can handle changes effectively and efficiently, minimizing any negative impacts on the project.

Task 11: Plan and manage procurement

16. During the project planning phase, the project team identifies that they will need to procure additional resources, such as hardware and software, to complete the project. They estimate the resource requirements and needs, communicate the requirements to the project sponsor, and develop a plan to manage the procurement process. What is the purpose of managing suppliers/contracts in a media campaign project?

a. To reduce project costs in order to satisfy stakeholders financial concerns

b. To identify project risks to avoid any positive risks that may occur

c. To ensure that project deliverables meet the required quality standards

d. To accelerate the project timeline in order to satisfy stakeholders

Rationale: The correct answer is (c) - to ensure that project deliverables meet the required quality standards. Managing suppliers/contracts in a media campaign project ensures that the project team can work with high-quality suppliers and vendors to deliver high-quality project deliverables that meet the required quality standards.

Task 12: Manage project artifacts

17. PM Anonymous, a project management consulting firm, is managing a large-sized Agile IT project for a client. The project team has been efficiently trained on all tools and systems to work in an Agile environment. Halfway through the project, the Scrum Master realizes

that some of the project artifacts are not up to date. Some team members have not been updating the Agile project management tool regularly, and there is confusion about the latest version of the code. What is the best course of action for the Scrum Master to manage this issue with project artifacts?

a. Ignore the issue and focus on completing the project on time

b. Schedule a team meeting to discuss the importance of updating the project artifacts

c. Train team members on how to use the Agile project management tool effectively

d. Implement a peer review process to ensure that all project artifacts are up to date

Rationale: The team has already been trained, so the correct answer is (b) - schedule a team meeting to discuss the importance of updating the project artifacts. Managing project artifacts is essential in Agile projects, as it ensures that all team members have access to the latest project information and can work collaboratively towards project goals. Scheduling a team meeting to discuss the importance of updating the project artifacts can help the Scrum Master address the issue and prevent similar issues from arising in the future. This enables the project team to work more efficiently, improve project quality, and deliver high-quality project deliverables on time and within budget.

Task 13: Determine appropriate project methodology/methods and practices

18. During the project initiation phase, the project manager meets with the client to discuss the project requirements and determine the appropriate project methodology. The client is interested in an Agile approach to project management, but the project team is more familiar with Waterfall methodology. The project manager considers the project's complexity, risk factors, and resource availability to determine the most appropriate project methodology for the project. After conducting a detailed analysis, the project manager recommends a hybrid project management methodology that combines elements of Agile and Waterfall methodologies. The project manager explains the hybrid methodology to the client and the project team and obtains their buy-in. What is the primary benefit of using a hybrid project management methodology for this project?

a. Enables the project team to use Agile practices to manage project uncertainty and complexity

b. Ensures that the project is completed within the established timeline and budget

c. Provides a clear and well-defined project plan with detailed deliverables and timelines

d. Enables the project team to use iterative and incremental development practices to improve project quality

Rationale: The correct answer is (a) - enables the project team to use Agile practices to manage project uncertainty and complexity. Using a hybrid project management methodology for this project allows the project team to leverage the strengths of both Agile and Waterfall methodologies. This enables the project team to use Agile practices to manage project uncertainty and complexity, such as conducting frequent reviews and adapting to changes quickly, while also ensuring that the project is completed within the established timeline and budget. This enables the project team to deliver high-quality project deliverables on time and within budget while meeting the customer's requirements

Task 14: Establish project governance structure

19. What is the purpose of defining escalation paths and thresholds in a large-sized Agile IT project?

a. To ensure that project information is properly stored

b. To ensure that project information is secure

c. To define how and when project issues will be escalated

d. To ensure that project deliverables meet the required quality standards

Rationale: The correct answer is (c) - to define how and when project issues will be escalated. Defining escalation paths and thresholds in a large-sized Agile IT project ensures that the project team knows how and when project issues will be escalated, ensuring that issues are handled effectively and efficiently.

Task 14: Establish project governance structure

20. PM Anonymous, a project management consulting firm, is managing a new project for a client. The project involves developing a new software application for the client's business. The project has a duration of twelve months, and the budget is $2,000,000. The project team includes ten developers, three business analysts, and one project manager. During the project planning phase, the project manager is tasked with defining the appropriate governance structure for the project and determining escalation paths and thresholds. What is the primary purpose of establishing a project governance structure?

a. To ensure that the project is completed within the established timeline and budget

b. To provide a clear and well-defined project plan with detailed deliverables and timelines

c. To ensure that the project is executed effectively and efficiently and that high-quality project deliverables are produced

d. To establish communication methods, channels, frequency, and level of detail for all stakeholders

Rationale: The correct answer is (c) - to ensure that the project is executed effectively and efficiently and that high-quality project deliverables are produced. Establishing a project governance structure is an essential part of project management, as it ensures that the project is executed effectively and efficiently and that the project team delivers high-quality project deliverables on time and within budget. The project governance structure defines the roles and responsibilities of the project team and stakeholders, as well as the communication and escalation paths, which are critical to ensuring that the project is successful. This enables the project manager to manage the project effectively & efficiently while ensuring that the project meets the customer's requirements.

Task 15: Manage project issues

> 21. What is the purpose of collaborating with relevant stakeholders on the approach to resolve project issues in a large-sized Agile IT project?

a. To ensure that project information is properly stored

b. To ensure that project information is secure

c. To identify areas for improvement and adjust the project as necessary

d. To address project issues in a collaborative and effective manner

Rationale: The correct answer is (d) - to address project issues in a collaborative and effective manner. Collaborating with relevant stakeholders on the approach to resolve project issues in a large-sized Agile IT project ensures that project issues are handled in a collaborative and effective manner, ensuring that the project stays on track and meets its objectives.

Task 15: Manage project issues

> 22. During the project execution phase, the project team encounters an issue with the software application's user interface. The client has raised concerns about the user interface, and the project team is struggling to address the issue. The project managers convene a meeting with the project team and stakeholders to discuss the issue and

identify the root cause. After analyzing the issue, the project team realizes that the user interface issue is related to a coding error in the software application. The project team determines that they need to rewrite a significant portion of the code to address the issue. What is the best course of action for the project managers to manage this issue?

a. Ignore the issue and continue with the project as planned

b. Assign blame to the project team member responsible for the coding error

c. Schedule an emergency meeting with the client to discuss the issue and potential solutions

d. Collaborate with the project team to develop a plan to address the issue and communicate the plan to the client

Rationale: The correct answer is (d) - collaborate with the project team to develop a plan to address the issue and communicate the plan to the client. Managing project issues is a critical task in project management, as it allows project managers to address project problems and prevent them from derailing the project. In this scenario, the project managers need to work collaboratively with the project team to develop a plan to address the user interface issue and communicate the plan to the client. This enables the project team to address the issue effectively, ensure that the project stays on track, and meets the customer's requirements. Additionally, this fosters open communication and collaboration with the client, which is essential to building and maintaining strong relationships and ensuring project success.

Task 16: Ensure knowledge transfer for project continuity

23. What is the purpose of confirming the approach for knowledge transfers in a large-sized Agile IT project?

a. To ensure that project information is properly stored

b. To ensure that project information is secure

c. To ensure that project knowledge is transferred effectively and efficiently

d. To identify areas for improvement and adjust the project as necessary

Rationale: The correct answer is (c) - to ensure that project knowledge is transferred effectively and efficiently. Confirming the approach for knowledge transfers in a large-sized Agile IT project ensures that project knowledge is transferred effectively and efficiently, ensuring project continuity and enabling the project team to meet its objectives.

Task 16: Ensure knowledge transfer for project continuity

24. What is the main benefit of conducting a knowledge transfer session at the end of a project?

a. Helps the project team members improve their skills and knowledge

b. Facilitates the transfer of knowledge from the departing team members to the remaining team members

c. Enables the project manager to evaluate the performance of the project team members

d. Provides an opportunity to celebrate the successful completion of the project

Rationale: The correct answer is (b) - facilitates the transfer of knowledge from the departing team members to the remaining team members. Conducting a knowledge transfer session at the end of a project is essential to ensure project continuity and prevent any knowledge gaps. In this scenario, the project manager arranges for a knowledge transfer session to transfer knowledge from the departing project team members to the remaining team members. This enables the remaining team members to acquire the knowledge and skills necessary to maintain and upgrade the inventory management system in the future. By facilitating the transfer of knowledge, the project manager ensures that the project's success continues, even after the project has ended.

Task 17: Plan and manage project/phase closure or transitions

25. What is the purpose of concluding activities to close out a large-sized Agile IT project, such as final lessons learned and retrospective?

a. To ensure that the project team has completed all the required tasks

b. To assess project performance and identify areas for improvement

c. To accelerate the project timeline in order to complete the scope

d. To reduce the project budget in order to return money to the cost-centers

Rationale: The correct answer is (b) - to assess project performance and identify areas for improvement. Concluding activities to close out a large-sized Agile IT project, such as final lessons learned and retrospective, allows the project team to assess project performance, identify areas for improvement, and apply these lessons learned to future projects. This enables the project team to continuously improve their processes and deliver high-quality projects.

Task 17: Plan and manage project/phase closure or transitions

26. The project manager meets with the project team to discuss the closure plan and identify the specific tasks that need to be completed. The team determines that they need to conduct a final project review to evaluate the project's success, identify any areas for improvement, and document the lessons learned. They also need to archive the project documents and transfer the final deliverables to the client. What is the main benefit of conducting a final project review?

a. To evaluate the performance of the project team members and give appropriate rewards and recognition

b. To identify areas for improvement and apply the lessons learned to future projects

c. To celebrate the successful completion of the project

d. To determine if the project met the customer's requirements and have a final commendation from the stakeholders

Rationale: The correct answer is (b) - to identify areas for improvement and apply the lessons learned to future projects. Conducting a final project review is an essential part of the project closure process. In this scenario, the project team conducts a final review of the project to evaluate its success, identify areas for improvement, and document the lessons learned. By doing this, they can apply the lessons learned to future projects, which improves project performance and reduces the risk of making the same mistakes in the future. Additionally, it enables the team to celebrate their successes and provide closure to the project.

PMP® BUSINESS QUESTIONS

1. As a project manager for a software development company, you have been tasked with developing a plan for a new project management software to be used within your organization. Which of the following is NOT a compliance category that should be classified during the project planning process?

a. Health and safety compliance

b. Environmental compliance

c. Quality compliance

d. Design compliance

2. You have assembled a team of developers, designers, and other stakeholders to work planning compliance for a project. Which of the following is NOT a method that can be used to support compliance during a project?

a. Peer review

b. Auditing

c. Training and education

d. Risk avoidance

3. Susan is a project manager for a construction project. She has identified certain safety measures that need to be implemented to ensure the safety of workers on the construction site. However, the team members are not following the safety measures and are taking shortcuts to save time and effort. What is the first step that Susan should take to address noncompliance in this project?

a. Analyze the consequences of noncompliance

b. Determine the necessary approach and action to address compliance needs

c. Classify the compliance categories

d. Measure the extent to which the project is in compliance

4. During the investigation, it is discovered that the team members had ignored the safety measures. Susan is frustrated and realizes that noncompliance is a serious issue that needs to be addressed immediately. Which of the following is NOT a potential threat to compliance that a project manager should have considered during project planning?

a. Lack of stakeholder engagement

b. Insufficient resources

c. Inadequate training and education

d. Excessive regulatory oversight

5. During acceptance testing of a deliverable in an ongoing project, a senior stakeholder complains that a functionality essential to their department was not addressed. What should the project manager do next?

a. Reopen the project and address the missing functionality

b. Address the missing functionality in the next project phase

c. Determine if the missing functionality was required and if it was included in the scope

d. Explain that the functionality was not included in the scope and cannot be addressed

6. What is the purpose of verifying that a benefit measurement system is in place to track benefits during a project?

a. To ensure that all benefits are identified and documented

b. To determine who will be responsible for realizing the benefits

c. To establish a process for tracking the progress of the project

d. To evaluate the effectiveness of the project in delivering value

7. Which of the following is NOT a step that a project manager should take to evaluate delivery options and demonstrate value?

a. Analyze the benefits and costs of each delivery option

b. Identify potential risk strategies associated with each delivery option

c. Select the option that is most cost-effective and implement it

d. Develop a plan for monitoring and measuring benefits realization

8. Why is it important to document agreement on ownership for ongoing benefit realization?
a. To establish a process for tracking the progress of the project
b. To determine who will be responsible for realizing the benefits
c. To identify potential risks and mitigation strategies for each delivery option
d. To select the option that provides the most value

9. What is the purpose of surveying changes to the external business environment during a project?
a. To identify potential risks to the project
b. To assess the impact of changes on the project scope/backlog
c. To identify opportunities for the project
d. To prioritize changes to the project scope/backlog

10. How should a project manager prioritize changes to the project scope/backlog based on changes in the external business environment?
a. By their potential impact on the project budget
b. By their potential impact on the project timeline
c. By their potential impact on the project's strategic objectives
d. By their potential impact on the project's technical requirements

11. What is the first action the project manager should take after being informed that work has stopped in one city due to safety issues, and authorities will not allow further project work for three months? a. Continue the project work in other cities as planned
b. Immediately halt all project work in all cities until the investigation is complete
c. Notify the authorities that the project work will continue as planned

d. Review the project plan and make necessary adjustments to accommodate the delay in one city

12. What steps should the project manager take to address the delay in one city and ensure the project remains on track?

a. Wait for the authorities to complete their investigation before taking any action

b. Reassign resources from other cities to make up for the delay in one city

c. Work with the team leader to identify the cause of the safety issue and implement necessary changes to prevent further delays

d. Inform the stakeholders that the project will be delayed in one city and that the project plan has been adjusted to accommodate the delay

13. What communication strategy should the project manager employ to ensure all stakeholders are informed of the delay in one city and the necessary adjustments to the project plan?

a. Only inform stakeholders who are directly impacted by the delay in one city

b. Wait until the investigation is complete before informing stakeholders of the delay

c. Clearly communicate the delay and necessary adjustments to the entire project team and all stakeholders

d. Leave the decision up to the team leader in the affected city to communicate the delay to local stakeholders

14. What risk management strategy should the project manager employ to prevent similar safety issues in other cities?

a. Wait until the investigation is complete before taking any action to address the safety issue

b. Implement additional safety measures in all cities to prevent similar issues from occurring

c. Conduct a risk assessment to identify potential safety issues in other cities and implement necessary changes

d. Assign a safety officer to oversee safety measures in all cities throughout the project

15. What is the primary goal of change management in a project?

    a. To prevent any changes to the project scope

    b. To minimize the impact of changes on the project schedule and budget

    c. To ensure that any changes to the project are properly controlled and communicated

    d. To encourage stakeholders to propose more changes to the project

16. Phill is a project manager working on a munitions clean-up project with the National Defense Department in Northern Arizona, where WW2 munitions were not properly disposed of. He has several contractors working with him on the project on a per-diem basis with very rigorous information checks, people-validation (before and during the project), health and safety checks from a third-party. Which of the following compliance categories must Phill consider in the PLI Agile project?

    a. Environmental compliance

    b. HR compliance

    c. IT compliance

    d. All of the above

17. Which of the following methods can Phill use to support compliance in the PLI Agile project?

    a. Risk assessment

    b. Legal action

    c. Health and safety inspections

    d. All of the above

18. Phill is a project manager working on a munitions clean-up project with the National Defense Department in Northern Arizona, where WW2 munitions were not properly disposed of. He has several contractors working with him on the project on a per-diem basis with very rigorous information checks, people-validation (before and during the project), health and safety checks from a third-party. What is one key step Phill should take to evaluate and deliver project benefits and value for the PLI Agile project?

    a. Assess organizational culture

    b. Verify measurement system

    c. Determine potential threats to the project

    d. Use methods to support risk compliance

19. Phill is a project manager working on a munitions clean-up project with the National Defense Department in Northern Arizona, where WW2 munitions were not properly disposed of. He has several contractors working with him on the project on a per-diem basis with very rigorous information checks, people-validation (before and during the project), health and safety checks from a third-party. Which of the following is a step Phill can take to evaluate and address external business environment changes for impact on scope in the PLI Agile project?

    a. Continually review external business environment for impacts on project scope/backlog

    b. Assess and prioritize impact on project scope/backlog based on changes in internal business environment

    c. Determine potential threats to compliance and schedule risks

    d. Use methods to support risk management activities

20. Phill is a project manager working on a munitions clean-up project with the National Defense Department in Northern Arizona, where WW2 munitions were not properly disposed of. Which of the following is a step Phill can take to support organizational change in the PLI Agile project?

a. Assess organizational revenues and strategy

b. Evaluate impact of organizational change to project and determine required actions

c. Verify measurement system is in place

d. Determine potential threats to remaining in scope and on schedule

21. Phill is the project manager on an Agile project at PLI. What is another step Phill can take to support organizational change in the PLI Agile project?

a. Determine potential threats to compliance

b. Use methods to support compliance

c. Evaluate impact of the project to the organization and determine required actions

d. Verify the project authorization and sign the project charter if there is none.

PMP® BUSINESS ANSWERS

1. As a project manager for a software development company, you have been tasked with developing a plan for a new project management software to be used within your organization. Which of the following is NOT a compliance category that should be classified during the project planning process?
   a. Health and safety compliance
   b. Environmental compliance
   c. Quality compliance
   d. Design compliance

Answer: d. Design compliance
Rationale: Classifying compliance categories is an important step in planning and managing project compliance. Health and safety, environmental, and quality compliance are common categories, while design compliance is not typically considered a compliance category.

2. You have assembled a team of developers, designers, and other stakeholders to work planning compliance for a project. Which of the following is NOT a method that can be used to support compliance during a project?
   a. Peer review
   b. Auditing
   c. Training and education
   d. Risk avoidance

Answer: d. Risk avoidance
Rationale: Methods to support compliance include peer review, auditing, training and education, and risk mitigation, but not risk avoidance, which is not a method for ensuring compliance. A team cannot logically avoid compliance, so this is the best response. All the other approaches are some form of mitigation.

3. Susan is a project manager for a construction project. She has identified certain safety measures that need to be implemented to ensure the safety

of workers on the construction site. However, the team members are not following the safety measures and are taking shortcuts to save time and effort. What is the first step that Susan should take to address noncompliance in this project?

a. Analyze the consequences of noncompliance

b. Determine the necessary approach and action to address compliance needs

c. Classify the compliance categories

d. Measure the extent to which the project is in compliance

Answer: b. Determine the necessary approach and action to address compliance needs

Rationale: The first step to address noncompliance in a project is to determine the necessary approach and action to address compliance needs. Analyzing the consequences of noncompliance, classifying compliance categories, and measuring the extent to which the project is in compliance are later steps.

4. During the investigation, it is discovered that the team members had ignored the safety measures. Susan is frustrated and realizes that noncompliance is a serious issue that needs to be addressed immediately. Which of the following is NOT a potential threat to compliance that a project manager should have considered during project planning?

a. Lack of stakeholder engagement

b. Insufficient resources

c. Inadequate training and education

d. Excessive regulatory oversight

Answer: a. Lack of stakeholder engagement

Rationale: Potential threats to compliance that a project manager should consider during project planning include insufficient resources, inadequate training and education, and excessive regulatory oversight, but not lack of stakeholder engagement, which is not a threat to compliance. This is the least relevant option to upholding compliance.

5. During acceptance testing of a deliverable in an ongoing project, a senior stakeholder complains that a functionality essential to their department was not addressed. What should the project manager do next?

a. Reopen the project and address the missing functionality

b. Address the missing functionality in the next project phase

c. Determine if the missing functionality was required and if it was included in the scope

d. Explain that the functionality was not included in the scope and cannot be addressed

Answer: c. Determine if the missing functionality was required and if it was included in the scope
Rationale: To avoid similar issues in the future, the project manager should determine whether the missing functionality was actually required and whether it was included in the scope. Reopening the project is not necessary as the project is ongoing. Addressing the missing functionality in the next project phase does not immediately address the problem and explaining that the functionality was not included in the scope may not resolve the issue.

6.   What is the purpose of verifying that a benefit measurement system is in place to track benefits during a project?

a. To ensure that all benefits are identified and documented

b. To determine who will be responsible for realizing the benefits

c. To establish a process for tracking the progress of the project

d. To evaluate the effectiveness of the project in delivering value

Answer: d. To evaluate the effectiveness of the project in delivering value
Rationale: Verifying that a measurement system is in place to track benefits during a project is important to evaluate the effectiveness of the project in delivering value. This helps project managers to track the progress of benefits realization and make necessary adjustments to ensure that project goals and objectives are being met.

7.   Which of the following is NOT a step that a project manager should take to evaluate delivery options and demonstrate value?

a. Analyze the benefits and costs of each delivery option

b. Identify potential risk strategies associated with each delivery option

c. Select the option that is most cost-effective and implement it

d. Develop a plan for monitoring and measuring benefits realization

Answer: c. Select the option that is most cost-effective and implement it

Rationale: Project managers should analyze the benefits and costs of each delivery option, identify potential risks and mitigation strategies for each option, and develop a plan for monitoring and measuring benefits realization. Simply selecting the most cost-effective option may not always be the best option in terms of value delivery.

8. Why is it important to document agreement on ownership for ongoing benefit realization?
a. To establish a process for tracking the progress of the project
b. To determine who will be responsible for realizing the benefits
c. To identify potential risks and mitigation strategies for each delivery option
d. To select the option that provides the most value

Answer: b. To determine who will be responsible for realizing the benefits
Rationale: Documenting agreement on ownership for ongoing benefit realization is important to determine who will be responsible for realizing the benefits. This helps project managers to ensure that the benefits identified in the project plan are actually delivered and that the project provides value to the organization. By doing this, the stakeholders and project team members can be held accountable for the successful realization of project benefits.

9. What is the purpose of surveying changes to the external business environment during a project?
a. To identify unknown risks to the project
b. To assess the impact of changes on the project scope/backlog
c. To identify financial sources for the project
d. To prioritize changes to the sprint backlog

Answer: b. To assess the impact of changes on the project scope/backlog
Rationale: Surveying changes to the external business environment during a project is important to assess the impact of these changes on the project scope/backlog. This helps project managers to determine the appropriate course of action and make necessary adjustments to ensure that the project objectives are being met.

10. How should a project manager prioritize changes to the project scope/backlog based on changes in the external business environment?

a. By their potential impact on the project budget

b. By their potential impact on the project timeline

c. By their potential impact on the project's strategic objectives

d. By their potential impact on the project's technical requirements

Answer: c. By their potential impact on the project's strategic objectives
Rationale: Prioritizing changes to the project scope/backlog based on changes in the external business environment should be done by their potential impact on the project's strategic objectives. This helps project managers to ensure that the project is aligned with the organization's overall strategic goals and that the changes made are the most beneficial for the project.

11. What is the first action the project manager should take after being informed that work has stopped in one city due to safety issues, and authorities will not allow further project work for three months? a. Continue the project work in other cities as planned

    b. Immediately halt all project work in all cities until the investigation is complete

    c. Notify the authorities that the project work will continue as planned

    d. Review the project plan and make necessary adjustments to accommodate the delay in one city

Answer: d. Review the project plan and make necessary adjustments to accommodate the delay in one city.
Rationale: After being informed that work has stopped in one city due to safety issues, and authorities will not allow further project work for three months, the project manager should review the project plan and make necessary adjustments to accommodate the delay in one city. This could include rescheduling resources, adjusting timelines, or identifying alternative solutions to ensure that the project can still be delivered as planned.

12. What steps should the project manager take to address the delay in one city and ensure the project remains on track?

    a. Wait for the authorities to complete their investigation before taking any action

    b. Reassign resources from other cities to make up for the delay in one city

c. Work with the team leader to identify the cause of the safety issue and implement necessary changes to prevent further delays

d. Inform the stakeholders that the project will be delayed in one city and that the project plan has been adjusted to accommodate the delay

Answer: c. Work with the team leader to identify the cause of the safety issue and implement necessary changes to prevent further delays.

Rationale: To address the delay in one city and ensure the project remains on track, the project manager should work with the team leader to identify the cause of the safety issue and implement necessary changes to prevent further delays. This could involve making changes to the project plan or making changes to the safety procedures to ensure that work can resume safely in the affected city.

13. What communication strategy should the project manager employ to ensure all stakeholders are informed of the delay in one city and the necessary adjustments to the project plan?

   a. Only inform stakeholders who are directly impacted by the delay in one city

   b. Wait until the investigation is complete before informing stakeholders of the delay

   c. Clearly communicate the delay and necessary adjustments to the entire project team and all stakeholders

   d. Leave the decision up to the team leader in the affected city to communicate the delay to local stakeholders

Answer: c. Clearly communicate the delay and necessary adjustments to the entire project team and all stakeholders.

Rationale: When there is a delay in one city due to safety issues, it is important for the project manager to communicate the delay and necessary adjustments to the entire project team and all stakeholders. This ensures that everyone is aware of the delay and any necessary changes to the project plan, which helps to maintain transparency and minimize confusion.

14. What risk management strategy should the project manager employ to prevent similar safety issues in other cities?

a. Wait until the investigation is complete before taking any action to address the safety issue

b. Implement additional safety measures in all cities to prevent similar issues from occurring

c. Conduct a risk assessment to identify potential safety issues in other cities and implement necessary changes

d. Assign a safety officer to oversee safety measures in all cities throughout the project

Answer: c. Conduct a risk assessment to identify potential safety issues in other cities and implement necessary changes. Action should be proactively taken.
Rationale: To prevent similar safety issues in other cities, the project manager should conduct a risk
assessment to identify potential safety issues and implement necessary changes. This helps to proactively identify and address potential safety issues, which minimizes the risk of future safety-related delays and helps to ensure that the project can be completed on time and within budget.

15. What is the primary goal of change management in a project?

a. To prevent any changes to the project scope

b. To minimize the impact of changes on the project schedule and budget

c. To ensure that any changes to the project are properly controlled and communicated

d. To encourage stakeholders to propose more changes to the project

Answer: c. To ensure that any changes to the project are properly controlled and communicated
Rationale: Change management is a process that is used to manage and control changes to a project. Its primary goal is to ensure that any changes to the project are properly controlled and communicated to all stakeholders. This helps to minimize the impact of changes on the project schedule and budget, and also ensures that any changes that are made are consistent with the project's objectives and goals. Preventing changes to the project scope is not the primary goal of change management, as changes to the project scope may be necessary for the project's success. Encouraging stakeholders to propose more changes to the project is also not the primary goal of change

management, as the goal is to ensure that any changes that are made are properly controlled and communicated, not to encourage additional changes.

16. Phill is a project manager working on a munitions clean-up project with the National Defense Department in Northern Arizona, where WW2 munitions were not properly disposed of. He has several contractors working with him on the project on a per-diem basis with very rigorous information checks, people-validation (before and during the project), health and safety checks from a third-party. Which of the following compliance categories must Phill consider in the PLI Agile project?
    a. Environmental compliance
    b. HR compliance
    c. IT compliance
    d. All of the above

Answer: d. All of the above
Rationale: Compliance categories such as environmental, HR, and IT compliance must be considered in any project, including an Agile project like the one being managed by Phill at PLI.

17. Which of the following methods can Phill use to support compliance in the PLI Agile project?
    a. Risk assessment
    b. Legal action
    c. Health and safety inspections
    d. All of the above

Answer: a. Risk assessment
Rationale: Methods such as risk assessment can be used to support compliance in a project, including an Agile project like the one being managed by Phill at PLI.

18. Phill is a project manager working on a munitions clean-up project with the National Defense Department in Northern Arizona, where WW2 munitions were not properly disposed of. He has several contractors working with him on the project on a per-diem basis with very rigorous information checks, people-validation (before and during the project), health and safety checks

from a third-party. What is one key step Phill should take to evaluate and deliver project benefits and value for the PLI Agile project?

a. Assess organizational culture

b. Verify measurement system

c. Determine potential threats to the project

d. Use methods to support risk compliance

Answer: b. Verify measurement system

Rationale: To evaluate and deliver project benefits and value, Phill should ensure that a measurement system is in place to track benefits.

19. Phill is a project manager working on a munitions clean-up project with the National Defense Department in Northern Arizona, where WW2 munitions were not properly disposed of. He has several contractors working with him on the project on a per-diem basis with very rigorous information checks, people-validation (before and during the project), health and safety checks from a third-party. Which of the following is a step Phill can take to evaluate and address external business environment changes for impact on scope in the PLI Agile project?

a. Continually review external business environment for impacts on project scope/backlog

b. Assess and prioritize impact on project scope/backlog based on changes in internal business environment

c. Determine potential threats to compliance and schedule risks

d. Use methods to support risk management activities

Answer: a. Continually review external business environment for impacts on project scope/backlog

Rationale: To evaluate and address external business environment changes for impact on scope, Phill should continually review the external business environment for any changes that could impact the project scope/backlog.

20. Phill is a project manager working on a munitions clean-up project with the National Defense Department in Northern Arizona, where WW2 munitions were not properly disposed of. Which of the following is a step Phill can take to support organizational change in the PLI Agile project?

a. Assess organizational revenues and strategy

b. Evaluate impact of organizational change to project and determine required actions

c. Verify measurement system is in place

d. Determine potential threats to remaining in scope and on schedule

Answer: b. Evaluate impact of organizational change to project and determine required actions

Rationale: To support organizational change, Phill should evaluate the impact of the change to the project and determine the necessary actions.

21. Phill is the project manager on an Agile project at PLI. What is another step Phill can take to support organizational change in the PLI Agile project?

    a. Determine potential threats to compliance

    b. Use methods to support compliance

    c. Evaluate impact of the project to the organization and determine required actions

    d. Verify the project authorization and sign the project charter if there is none.

Answer: c. Evaluate impact of the project to the organization and determine required actions

Rationale: To support organizational change, Phill should also evaluate the impact of the project to the organization and determine the necessary actions to take in response to that impact. This will help ensure that the change is successfully integrated into the organization and that the project can deliver the expected benefits.

About the Author

Phill C. Akinwale, PMP has managed operational endeavors, projects and project controls across government and private sectors in various companies, including Motorola, Honeywell, Emerson, Skillsoft, Citigroup, Iron Mountain, Brown and Caldwell, US Airways and CVS Caremark. With his extensive experience in various facets of Project Management and rigorous project controls, he has trained project management worldwide (NASA, FBI, USAF, USACE, US Army, Department of Transport) across five PMBOK® Guide editions over the last 15 years.

He holds twelve project management certifications with six in Agile Project Management (CSM, PMI-ACP, PSM, PSPO, PAL, SPS). As a John Maxwell Certified Coach and Speaker, Phill delivers workshops, seminars, keynote speaking, and coaching in leadership and soft skills. Working together with you and your team or organization, he will guide you in the desired direction and equip you to reach your goals. Books he has authored include: The No-Good Leader, Earned Value Basics and Project Management Mid-Level to C-Level.